PHILOPONUS
On Aristotle On Coming-to-Be and Perishing 1.6-2.4

PHILOPONUS
On Aristotle On Coming-to-Be and Perishing 1.6-2.4

Translated by C.J.F. Williams

Introduction by Sylvia Berryman

B L O O M S B U R Y
LONDON • NEW DELHI • NEW YORK • SYDNEY

Bloomsbury Academic
An imprint of Bloomsbury Publishing Plc

50 Bedford Square
London
WC1B 3DP
UK

1385 Broadway
New York
NY 10018
USA

www.bloomsbury.com

First published in 2001 by Gerald Duckworth & Co. Ltd.
Paperback edition first published 2014

British Library Cataloguing-in-Publication Data
A catalogue record for this book is available from the British Library.

ISBN HB: 978-0-7156-2854-6
PB: 978-1-7809-3877-6
ePDF: 978-1-7809-3876-9

Acknowledgements
The present translations have been made possible by generous and imaginative funding from the following sources: the National Endowment for the Humanities, Division of Research Programs, an independent federal agency of the USA; the Leverhulme Trust; the British Academy; the Jowett Copyright Trustees; the Royal Society (UK); Centro Internazionale A. Beltrame di Storia dello Spazio e del Tempo (Padua); Mario Mignucci; Liverpool University; the Leventis Foundation; the Humanities Research Board of the British Academy; the Esmée Fairbairn Charitable Trust; the Henry Brown Trust; Mr and Mrs N. Egon; The Netherlands Organisation for Scientific Research (NWO/GW).

Typeset by Ray Davies

Contents

Preface *Richard Sorabji* vii
Introduction *Sylvia Berryman* 1
Textual Emendations 19

Translation 21
1.6. Contact 21
1.7. Acting and Being Affected (1). Like and Unlike 41
1.8. Acting and Being Affected (2). Empedocles and the Atomists 56
1.9. Acting and Being Affected (3) Actuality and Potentiality 88
1.10. Mixing 94
2.1. The Elements and Prime Matter 114
2.2. The Primary Contrarieties 123
2.3. The Contrarieties and the Elements 135
2.4. Reciprocal Transformation of the Elements 142

Notes 151
English-Greek Glossary 173
Greek-English Index 175
Subject Index 179

Preface

Richard Sorabji

It is a tribute to Christopher Williams that his translation was in such a good state when he died, even though he had not had time to revise it. Sylvia Berryman's introduction and notes bring out the many subjects of high interest in these chapters of Philoponus. I will only point to the recent identification by Emma Gannagé of a previously lost commentary by Philoponus' great predecessor Alexander, a commentary to which Philoponus often refers. Gannagé has found a medieval Arabic MS which confirms the conjecture that another MS contains an Arabic version of much of Alexander's commentary on Aristotle's *De Generatione et Corruptione* (*GC*) 2.2-6.[1] We plan to provide a translation when her edition is ready.

So far, she has drawn attention to two topics in particular. Alexander is concerned to deprive Aristotle's natural places of any causal power in producing the natural rise and fall of the four elements. This would entirely fit with Philoponus' own view in his *Physics* commentary at 632,4-634,2 and 581,18-31.[2] But Philoponus' *GC* commentary does not stress the matter, perhaps because of another point of Alexander's, that the heaviness and lightness of the elements is not related in *GC* 2.2 to their other qualities – hot, cold, fluid and dry – because heaviness and lightness are relevant only to motion, not to acting and being affected in other ways.

The other relevant point that becomes clear from Gannagé's discussion is how far Philoponus was indebted to Alexander for his treatment of the question why Aristotle left incomplete his attempt to reduce other qualities of the elements to the basic four – hot, cold, fluid and dry. Dense and rare are not mentioned, rough and smooth, heavy and light are not reduced. At 214,22ff. and 225,28-31, Philoponus offers several explanations, referring occasionally to Alexander. The Arabic shows that Alexander had supplied more of these explanations than would otherwise appear. One such explanation (translated from the Arabic version of Alexander on p. 51 of Gannagé's article) is that rough and smooth, like heavy and light, do not imply acting on other things and being affected by them. Another explanation, which Gannagé reports as offered by Alexander at 330b10, is repeated by Philoponus at 330b11. It is that dense and rare can be reduced to hot and cold. Thus we have new evidence on the importance of Alexander to Philoponus.

I wish to thank the following scholars for commenting on Christopher Williams' translation: David Furley, David Hahm, Alan Lacey, Gerard Naddaf and Gisela Striker. I also thank David Robertson for supplying the indexes and Sylvia Berryman for coordinating comments and supplying the introduction and most of the notes.

Notes

1. Emma Gannagé, 'Alexandre d'Aphrodise *In De generatione et corruptione apud* Gâbir b. Hayyân, K. al-Tasrîf', *Documenti e studi sulla tradizione filosofica medievale* 9, 1998, 35-86.

2. Translated and discussed by Richard Sorabji in *Matter Space and Motion*, London and Ithaca N.Y. 1988, 211-13.

Introduction

Sylvia Berryman

This volume contains the second part of C.J.F. Williams' translation of Philoponus' commentary on Aristotle's work *On Coming-to-Be and Perishing*.[1] When Williams translated Aristotle's work for the Clarendon Series in 1982, he acknowledged a debt to Philoponus as the most philosophically rewarding interpreter of Aristotle.[2] This volume shows Philoponus not only providing assistance in understanding Aristotle, but also developing his own solutions to the problems he encounters.

The chapters in this volume represent a change of topic from the varieties of change to the preconditions for change. In chapters 1.1-5 Aristotle had focused on distinguishing the four different varieties of change: coming-to-be and perishing, alteration, growth and diminution, local motion; that is, substantial, qualitative, quantitative and local change respectively. After summarizing these results and recalling the relevance of the distinction of kinds of change to Aristotle's criticism of his predecessors, Philoponus turns in the remainder of his commentary on Book 1 to three topics closely intertwined both with one another and with Aristotle's critique of Presocratic views on change. These are contact, mixing, and the reciprocal pair acting and being affected.

Aristotle rejects the idea that invisible microstructures like void spaces or passageways explain alteration: he interprets alteration as a process of becoming actually what was previously present in potential. Philoponus highlights two difficulties in Aristotle's account, developing original solutions to these. One concerns the sense in which components of a mixture are said to be present in potential. Like many modern commentators, Philoponus doubts that either Aristotle's first or second sense of potential can capture adequately the sense in which things remain themselves in potential while in a mixture: he boldly defines a third sense of potential.[3] A second problem concerns the relationship between a substrate and the properties that it possesses. Philoponus rejects a view that qualities are *produced by* a mixture of hot and cold, wet and dry, claiming that qualities rather *supervene on* suitable mixtures. That changes in a mixture will affect some but not others of its supervening qualities suggests that qualities both admit of degree and have various ranges of mixtures on which they can supervene.

Contact and causation

Aristotle's views on contact or touching – Williams uses either translation as appropriate – appear at their most developed in *Physics* 5, where Aristotle differentiates various degrees of cohesion: Philoponus refers to that discussion here. Contact is used to distinguish mutual interaction from nonreciprocal causing of change. Aristotle holds that acting and being affected involve mutual interaction between bodies; he differentiates this from causing change without being reciprocally affected.

Philoponus – and Alexander of Aphrodisias, the other commentator he cites – are eager to show that Aristotle intended his discussion to allow for cases like the divine unmoved mover, which causes change without itself being changed. They also take Aristotle's discussion to extend to the unchanging fifth element, which raises a problem of explaining why Aristotle says that things which affect one another have a common matter. Philoponus takes Aristotle to stress that a common matter is required for *mutual* interaction, in order to allow that acting might occur without reciprocal affect between things with no common matter, as the heavens move the sublunary spheres.[4] There is an extended sense of touching – here Philoponus reminds us of Aristotle's notion of focal meaning – in which the image of a desired object might be said to 'touch' us, although it is not itself touched by us. It is in this sense that the divine unmoved mover moves the heavens.

As contact, in its strict sense, is the hallmark of acting and being affected, it requires elucidation. Philoponus argues that contact in the strict sense must be confined to physical bodies: the sense of contact applicable to mathematical objects can only be notional, just as they exist only conceptually. Aristotle thinks that being in contact, strictly speaking, requires having position: he interprets position to mean 'being in place' and understands this to imply the existence of natural place. As this does not obviously follow from the definition of contact as 'having extremities together', Philoponus tries to explain this connection, and why the idea of being in place leads Aristotle to infer that things in place have, necessarily, properties of being above or below and hence are heavy or light.

As acting and being affected are defined in a way that implies having place, Philoponus sees the need to explain how the spheres of the heavens aren't affected by one another, despite having position and being in contact.[5] Rather than suppose that Aristotle's discussion isn't intended to apply to the heavens, Philoponus considers interpretations that might allow this. He suggests that the spheres are only in place as a whole, not relative to each other, and so don't act on one another in virtue of having place. In another controversy over whether the stars are to be included in a discussion of contact, Philoponus finds it necessary to refine Aristotle's discussion of contact so as to limit it to sublunary bodies.

It is not always clear whether the terms acting and being affected are used broadly to include all kinds of change, or whether the terms are employed in their more specific application to alteration only. Philoponus has a reason for distinguishing carefully between acting, in the narrow sense that implies reciprocal change, from causing change more broadly. He needs this to explain why Aristotle's God is not said to 'act' on the universe.

Aristotle had ascribed to a prime mover a specific role of causing local motion. Proclus and Syrianus recognized that Aristotle only makes the prime mover responsible for the motion of the cosmos, not for its existence. But Ammonius interpreted Aristotle's argument about infinite power to be implicitly arguing for an efficient, *poiêtikon*, cause for the existence of the cosmos. Simplicius reports and adopts this interpretation.[6] One possible objection to this attempt to interpret Aristotle as in agreement with Plato is that Aristotle calls the prime mover a final rather than an efficient cause.

Philoponus has a different strategy to reconcile Aristotle's views with those of Plato's *Timaeus*, which answers this objection. He admits that Aristotle *doesn't* say God is an efficient cause, but distinguishes this from creating the cosmos. Philoponus' argument focuses on the relationship of the adjective *poiêtikon*, used for Aristotelian efficient cause, to the verb *poiein*, 'to act'. While *poiein* can have a broader sense and can even mean to make something, Philoponus suggests that Aristotle confines its use to qualitative change, which implies mutual affection. It would thus be inappropriate for use of a divine creator, since it implies that the divine is changed. Philoponus prefers the term from Plato's *Timaeus*, 'to fashion' or 'create', *dêmiourgein*, to describe the divine role in creating the cosmos.[7] He also canvasses the description of God as a *proagôgos*, a 'procurer', i.e. that he gets others to act on its behalf. The point of this term again seems to be to avoid saying that God might 'act' in such a way as to be susceptible to mutual effect. The dispute is not so much about God's role, but how Aristotle's view can be made compatible with Plato's. Philoponus' strategy, in contrast to that of Simplicius and Ammonius, is to show that Aristotle's term for efficient cause, *poiêtikon*, is not appropriate for a creator.

In discussing whether Aristotle implicitly describes God as an agent or the efficient cause of the cosmos, Simplicius takes the definition of efficient cause to be that from which change has its origin.[8] Philoponus gives a different account of an efficient cause, characterizing it as something that brings about an alteration, bringing something to a form that is not yet completely present. By contrast, he seems to suggest that with a final cause the end and the disposition are already present. The distinction he has in mind seems to echo that between change, *kinêsis*, and activity, *energeia*: the one involves reaching a new state, the other is the actualization of a potential already present by removing any impediment.[9] The point is clearly that efficient causes seems to Philoponus to be inherently temporal and hence to imply something incomplete or imperfect.[10] Final cause seems to him more appropriate to divine activity.

Acting and being affected: like and unlike

Central to the account of alteration is the question of what things can act on and affect one another. Aristotle classifies Presocratic theories into those claiming action happens between likes and those claiming it happens between unlikes, classifying a multitude of heterogeneous theories in terms of a single issue.[11] The apparent conflict – that things acting on each other are both like and unlike – is resolved by the Aristotelian distinction into genus and species. Aristotle says that things acting on and affected by each other need to be like in genus but unlike in species. They need to be sufficiently similar as to lie on the same continuum: white can be made black but not sweet.

The reasons why likeness alone cannot define which things act require more careful elaboration: Philoponus has some considerations of his own to introduce. Aristotle's justification for the restriction against like acting on like is simply that things would then act on themselves and that this would create a problem in the argument for a first cause. His reasoning is unclear, as Philoponus notes. Philoponus recognizes the strength of Aristotle's point that if it were merely *qua* 'being like' that things act and are affected, things would seem able to act on themselves. He thinks the concern is that things might destroy themselves, so that nothing would be imperishable.[12] He considers whether this could be avoided if there were some things imperishable in their own nature.

Although Aristotle does think the danger of everything being destroyed requires the existence of something unchanging, he need not have meant that things able to act on themselves would thereby be able to *destroy* themselves. Aristotle is concerned that if things could *instigate* change, there would be no need for a sequence of causation leading back to a first cause. Philoponus later comes up with an argument against the idea that it could be in the nature of things to destroy themselves. It might also seem that Aristotle could have addressed any threat of self-destruction inherent in the principle that 'like acts on like' by a distinction between being like in kind and being the same in number. If he had simply stipulated that things which act on one another must be numerically different, no problem about things acting on themselves would arise. Philoponus reminds us that Aristotle's argument depends on the claim that it is *qua* being like in kind that things act on one another, as if to rule out this response.

But more than mere numerical difference is required for acting. Philoponus draws out some implications of the claim that things are not affected by something exactly the same in kind. If two things were exactly alike, neither would be any more inclined to act on the other; moreover, inasmuch as things act by making the other like itself, this would be impossible for things that are too much alike. Either possibility implies that acting involves transfer of form from agent to patient. This is a model of causation to which Philoponus brings some important restrictions.[13] He has earlier suggested that not every case of acting involves making the other like itself;[14] he later differentiates those contraries which are capable of acting from those which aren't.[15] Philoponus will talk of qualities or powers of different substances in a mixture restraining and overcoming each other, a different model of change from transfer of form.

Aristotle had used the claim that things acting and affected by one another must be like in genus to delimit the kinds of things that can interact. Philoponus shows that the kinds of qualities that can act on one another must be even more delimited than Aristotle suggested. Philoponus points out that not only must colour act on colour, but that the matter must be sufficiently alike: the black in ebony cannot act on the white in milk. As Philoponus thinks that in some sense all physical bodies have a common, prime matter, it would be vacuously true that all change of physical bodies takes place between things with a common matter:[16] the possible qualities a particular substance can acquire are limited by a more proximate matter. In saying that there must be a common matter as substrate of change, Philoponus introduces more specific restrictions on the kinds of changes that can occur.

This is one of the refinements Philoponus makes to Aristotle's account of qualitative change. There has been some debate about whether the contributions of the commentators amount to any real advance in clarity or precision regarding the natural world. Sambursky drew attention to the importance of the concept of suitability of matter, *epitêdeiotês*, calling it a mark of an 'increasing mechanical-mindedness' in later Greek thought.[17] Others have noted the particular role of this notion of suitability in Neoplatonist thought: Plotinus' notion of causes that are present everywhere means that the suitability of the receiver must be taken into account to explain why effects might be limited.[18] Sambursky suggests that the concept of suitability comes to be used, especially by Philoponus, in a more precise way than Aristotle's rather broad concept of *dunamis* or potential, to mean a sufficient condition for change.[19] Others have raised objections to Sambursky's view that the usage becomes technical.[20]

Where Philoponus uses 'suitability' as a name for Aristotelian first potential, it merely introduces a new name for a familiar distinction, and one for which Sambursky's characterization of it as a 'sufficient condition' is moreover too strong.[21] A more specific use occurs when Philoponus compares the fact that the black of ink can overpower the white of milk but the black of ebony is unable to: the density of the one kind of matter renders its qualities unsuitable to act on qualities in other matter. Philoponus may have specific ideas about physical limitations to causal interaction. He later endorses Aristotle's view that liquids are more able to affect each others' qualities than solids because they are 'more easily bounded' i.e. more divisible and hence among the substances more finely juxtaposed that act on each other more readily.[22]

Mixture and supervenience

Philoponus' views on the suitability of matter for form become clearer in the claim that qualities supervene on the matter. He distinguishes between the view that qualities are *produced* by the mixture of matter, as in some medical theories, and that qualities *supervene* from outside on suitable mixtures of hot, cold, wet and dry.[23] The idea that form comes from an external source has an obvious role in a theory that soul exists independent of the body;[24] Philoponus adopts this as an account of the relationship of qualities and forms generally to the material substrate. The same mixture will have, simultaneously, different qualities – honey is yellow and sweet – and moreover different mixtures also have these qualities. A change in the substrate need not produce a change in any given supervening quality; qualities vary independently of one another following changes in the substrate. Heating honey changes its colour but not its sweetness. Philoponus seems to take the fact that mixtures and qualities are not in unique correspondence to show that the mixture does not produce qualities. Aristotle introduces the idea that primary qualities in a mixture may not be in their extreme form but occur in a many intermediate degrees (*GC* 2.7, 334b8ff.). Philoponus develops the notion of a 'range' or latitude in the material conditions of qualities to provide some of the detail wanting from the notion of suitability alone.[25] A homoeomerous substance – say, flesh – is, Philoponus tells us, instantiated *within a certain range of variation* in the mixture of the component elements. The qualities that any given homoeomer can assume are those appropriate to that range of mixtures of components. Aristotle had recognized that one colour of flesh will change to another *colour* of flesh; Philoponus adds that it will also change to another colour of *flesh*. It's not that different colours are instantiated in the same mixture of matter, but rather that different mixtures, which take on different colours, can all be cases of flesh. Philoponus takes the common matter not to be the *particular* mixture, which would have a determinate colour, but the *range* of mixtures, any one of which could be flesh. Flesh doesn't take on the colour of ebony, but only one of the possible colours of flesh.

The background to the Neoplatonist usage of range of qualities may lie in the commentaries on the *Categories*. When Aristotle considers a distinction between the quality itself and the qualified thing, he considers a view that only the qualified thing admits of degree while the quality itself is never 'more or less' (*Cat.* 10b26-11a4).[26] Iamblichus objects that if there is a close correspondence between the quality and what is qualified, the quality ought to undergo the same affections as the qualified thing.[27] The idea that a quality undergoes intensification and relaxation – *epitasis* and *anesis* – is introduced in this regard. Philoponus accepts that a quality admits of different degrees and that these supervene on possible mixtures of the components.[28] It is worth enumerating the different ways in which the idea of range works in qualitative change.

(1) Restriction of qualities possible to a substance. The idea that there is a certain range of mixtures for a certain substance helps explain why the possible qualities are specific to the kind of substance. A colour change in flesh must be between colours that can be instantiated in the possible mixtures that can take the form of flesh; there is no change from the colour of flesh to that of pitch, because pitch black is not one of the colours that can be instantiated by the matter of flesh. The matter for qualities is not a particular mixture but a range of mixtures.

(2) Independent variation of qualities. Philoponus addresses the difficulties with the view that certain qualities either result from a mixture, as is commonly held in the medical schools, or 'supervene from outside on the suitability of the mixture'. A problem for both views is that if qualities of a homoeomerous substance are said to depend on a particular ratio of components, it does not seem reasonable that the qualities of that mixture could vary independently during change of the mixture: for honey to change colour, for instance, without losing its sweetness. Philoponus proposes that there might be different thresholds for the instantiation for different qualities. A certain mixture might just be over the limit for instantiating yellowness but well within the range for sweetness. So a small variation in the mixture could lead to change of colour but leave taste unaffected.

(3) Accidental change of a quality definitive of an element. If water is cold and wet by its nature, it might seem puzzling that water could be heated without thereby turning into another substance. This makes explicit what Aristotle in practice recognizes: that water can vary in temperature to a certain degree without undergoing substantial change. At one point Philoponus talks of the 'range of heat', suggesting that there is a distinction within the form of heat between the kind that is constitutive of elements and that which is merely an accident of an element. The use of 'range' here suggests that ordinary heat is different in degree rather than kind from that involved in elemental transformation.[29]

These all concern the suitability of mixtures to qualities. Although the terminology isn't as clearly parallel, De Haas has suggested that a number of Philoponus' views on quantity can be interpreted in terms of the same ideas of suitability and range.[30]

(4) Relative quantities of ingredients in a mixture. In contrast to the Stoics, Aristotle does not regard a single drop of wine as being mixed in an ocean of water. Rather, when the relative quantities of ingredients are not within a certain proportional range, what happens instead of mixture is that the lesser ingredient is overpowered and a case of growth occurs. The

explanation given by Philoponus is that mixing isn't a matter of juxtaposition of parts but of bodies acting on each other. To do so their relative quantities need to be proportionate, within a certain range, or their powers to act on one another are overcome.

(5) Restricting size limits during interelemental transformation. In commenting on chapter 1.5, Philoponus discusses the variation in size associated with the expansion and contraction of elements as they transform into one another. In his *Contra Proclum*, Philoponus notes that there must be a limit to the possible dimensions specific to the different elements that a given amount of matter can come to occupy.[31] Aristotle thinks that the same ratio must hold between the quantity of air produced from a small amount of water as between the entire body of air and the entire body of water, i.e. that the ratio is fixed.[32] Philoponus' attempt at *in GC* 1.5 to account for the fact that other bodies adapt to fill the space available when one case of elemental transformation takes place suggests that air especially can compress or rarefy within a certain range. There is some flexibility to the amount of space it can occupy without transforming, but extreme compression alone will turn it into water.

(6) Minima naturalia. This is a notion that Aristotle introduced in *Physics* 1.4 as part of his argument against Anaxagoras' notion of 'everything in everything', whereby a piece of bread contains morsels of flesh, bone and blood etc., a process continued indefinitely as these morsels in turn contain parts of every homoeomer. Aristotle argues from the fact that living beings only occur within a certain size range to the fact that the parts of living beings also only occur within a certain size range. He thinks homoeomers can't be indefinitely small. Philoponus takes up this idea in his commentary on the *Physics* especially, where he compares this idea to the claim that every form supervenes from outside and needs a suitable mixture: it must be suitable in quantity as well as relative proportions.[33] This has sometimes been taken as a kind of atomism: this doctrine isn't that flesh is composed of independently existing flesh parts of determinate size, merely that there is a smallest quantity of mixture on which the form of flesh can supervene.[34]

Philoponus seems to take minima naturalia to apply not just to parts of living beings, but also to elements. This is a significant extension of Aristotle's view. De Haas argues that the extension of minima naturalia to all forms, even those of the elements, is used to ground the view that there is a fixed quantitative range for interelemental transformation. In *Contra Proclum* this range of variation is credited to the three dimensional, which is that surviving all changes and having definite quantitative limits.[35] Philoponus clearly sees all this as grounded on the view that forms supervene on suitable mixtures of matter. He thinks that idea that there is a smallest quantity of matter on which a given form can supervene is distinct from the view of the atomists that there are smallest indivisible bodies. Every body is infinitely divisible as an extension, even if its form is not infinitely divisible.[36]

Supervenience and causation

One consequence of the views developed so far becomes apparent in the discussion of causation. Earlier it was mentioned that changes only occur between certain kinds of qualities, and that the suitability of the matter can restrict the kinds of things that can act on each other. But there are suggestions of a more profound rethinking of the whole nature of causation. Philoponus' rejection of the so-called 'transmission theory of causation' has been noted by Mourelatos;[37] it seems that the reasons for this rejection lie with the doctrine of supervenience.

When Philoponus realized that 'having a common matter' was insufficient restriction on the kinds of things that can act on one another, he recognized that qualities changing into one another must have mixtures within a certain range. It isn't just that qualities act on one another, but that they must have suitable matter. But Philoponus also considers some qualities to be efficacious rather than others. While white and black may change into one another, it seems to be the four qualities of the mixture on which the colour supervenes that do the actual causal work.

Philoponus' view on the reduction of qualitative change to the four primary qualities draws heavily on *Meteorology* Book 4, to which he refers. This programme of reducing causation to the four qualities seems to suggest that while forms supervene on mixtures of these four qualities, the causal account will reduce to changes in the matter. The supervening qualities are said to 'act on' one another in the sense of changing into one another; but the underlying primary qualities drive the change. What makes ice hard is cold, not hardness; hot and cold even more than wet and dry have the power to effect change.

Acting and being affected: microstructures

The above problems concern the relationship between qualities and matter. Philoponus' notion of qualities supervening on the matter is explicitly contrasted to the view that qualities are produced by the matter, but should also be distinguished from the view that qualities are merely appearances produced by the matter. Aristotle's criticism of Empedocles and the atomists together centres on the attempt to posit a material substructure which could alone explain macroscopic qualities. The difficulty, from an Aristotelian point of view, is that if qualitative change is supposed to be explained by rearrangement of smallest parts, it seems that the same question could be asked of the parts: how alteration then takes place once the parts are rearranged or juxtaposed. Saying that particles spread throughout the empty spaces or passageways of a body does not seem to show how the matter of the body itself alters. The force of Aristotle's criticism seems to depend upon whether a distinction of type between the microscopic and macroscopic structure can be sustained.

Philoponus' defence of Aristotle has as its background the argument about void and its relation to motion. Aristotle denies the atomist claim that change depends on void. Philoponus takes the opportunity to summarize arguments against the void, not only those about the divisibility of the atoms, but also those based on their infinite number and their impassivity. He draws on material from *On the Heavens* and the mathematical critique of indivisibles, as well as introducing a difficulty about perception.

Here the question about the usefulness of microstructures for explaining qualitative alteration recurs. Any perceptual awareness seems to involve contact. But for the atomists, contact must be either with a body or with void space, and both atoms and void are impassive. Atomist accounts could show how motions or effluences associated with perception enter the body but not how they are in fact perceived. Philoponus draws on Plato's *Theaetetus* for an argument that perception is impossible without being affected.

Besides atomism, the other account depending on microstructures is Empedocles' passageways. Aristotle had posed a dichotomy analogous to that facing the atomists, that change happens either by affecting the body through which the passages run, or else by affecting what is in the passageways themselves. The latter option would not explain how the body itself, outside of its passageways, were changed. But if the body outside the pores itself were changed by the body moving through the passageways, the passageways seem incidental to the explanation of change. Philoponus adds to this a third possibility, which he attributes to Plato. This is that the process of affecting takes place, neither in the body nor its passageways, but at the juncture of the two. This is also rejected.

Besides the undeniable visible passageways of the body like veins, which, as Philoponus notes, are indeed recognized by Aristotle, passageways are used freely in Book 4 of *Meteorology*. In the latter work, qualitative properties of different kinds of material – properties like fissibility – are attributed to the presence of passageways. This has struck some interpreters as sufficiently unAristotelian that they considered the work to be by someone else.[38] But the recognition that some qualities of materials might be due to a differential structure need not compromise the more fundamental supposition that qualities are present throughout matter and not essentially supervenient on the arrangement of smallest parts. Philoponus echoes Aristotle's suggestion that features like the seams in marble might be called *poroi*, meaning not empty channels, but simply points or lines at which the material is more easily affected.

Given Aristotle's apparent concessions towards accepting some kinds of passageways in matter, it is critical to show that, in cases like perception, passageways are not the primary means of explaining alteration and being affected. Some philosophers – including Plato – have been tempted to infer from the existence of passageways in the body leading back from the sense organs to the idea that perception depends on the movement of bodies through passages.[39] This is particularly evident in the idea that sight depends on a body analogous to light that passes through air. Aristotle rejects this idea wholeheartedly as an account either of perception or of light. Here Philoponus takes up the criticism of the view that sight depends on rays passing through a medium like air or water. Philoponus echoes criticisms of a visual ray theory of sight from his commentary on Aristotle's *De Anima*, particularly that this would give us a *pointilliste* and not a continuous image, and that if visual rays were bodies they would need to pass through one another to explain the ubiquity of visual images.[40]

As in the discussion of passageways, so too in the account of mixture the issue of the level of analysis arises. Aristotle rejected the idea that mixtures might at the microstructure be mere juxtaposition of parts, because this would imply that mixture is relative to perception, i.e. that a sharper-sighted observer would judge differently whether something was mixed or merely mingled. This would make the concept of mixture observer-relative, while, as Philoponus stresses, it should not be observer-dependent at all.

Potentiality and mixture

After summarizing the ground covered so far, Philoponus turns to the potential-actual distinction, which does most work in the ensuing discussion of mixture. Philoponus talks about different senses of 'potential', distinguishing that of being suitable, *epitêdeiotês*, from that of having a disposition or *hexis*.[41] He refers to Aristotle's notion of first and second potentiality in *De Anima*, where a key distinction is made between the sense in which a child is capable of doing geometry and an experienced geometer who is currently asleep is capable of doing geometry. The first kind of potentiality involves the suitability of the organism to acquire the respective disposition, an acquisition which would require a change. The second kind of potential is that of someone who can exercise the capacity without being further changed, once whatever is impeding its operation, such as sleep, is removed.

Philoponus, like many modern commentators, is dissatisfied with these two alternatives and suggests that the potentiality of an ingredient in a mixture must be a third kind of potential. He suggests that, like a drunk geometer, the components still retain their dispositions but are impeded in their functioning.[42] The components in a mixture exhibit restrained, *kekolasmenôs*, activity in contrast to its pure, *eilikrinês*, state. In contrast to the sleeping geometer, the drunk geometer couldn't get it right if he tried.[43]

It is not exactly clear whether Philoponus thinks of the idea that there is a third sense of potential as Aristotle's or his own. Frans de Haas notes that in the commentary on 1.10, it is unclear what is tempered in a mixture: whether ingredients exist in tempered form or whether it is the characteristic qualities that are said to be restrained in mixture. Moreover, if mixture is merely an alteration of qualities, there is a real danger of collapse into mcre alteration. De Haas suggests that *in GC* 1.10 should be regarded as attempting to present Aristotle's view, with all its difficulties, while in a later chapter, *in GC* 2.7, Philoponus formulates his own view. In the latter, elements do not persist in mixtures: it is the primary qualities that are present in tempered form.[44]

Philoponus tries to shore up Aristotle's account of several other factors affecting the ease and degree of mixture. He elaborates on Aristotle's explanation of why small quantities of material act on one another more quickly and why it is that the effect of qualities is not always – apparently – directly proportional to the quantity of matter. Other problems are to explain why fluids mix better, and why in some cases it isn't clear whether mixture or growth is taking place.

The elements and the contraries

The first four chapters of Book 2 are devoted to establishing the case for Aristotle's own view that there are four primary qualities composing the four elements. One question that needs to be settled is whether it is possible for anything other than the four elements to exist as an independent and separable substance. Here Philoponus takes on Aristotle's criticisms of Plato's 'receptacle' or 'omnirecipient' matter. One criticism Philoponus thinks can be partly answered concerns Plato's view that things should properly be characterized by the matter rather than the form: Philoponus thinks this is reasonable in the case of alteration, though not of coming-to-be. He also defends the *Timaeus* account that the elements are composed of plane figures from the charge that these are merely two-dimensional, claiming that Plato must mean the surfaces to be corporeal entities. Philoponus seems to accept the force of the criticism that corporeal bodies shouldn't be composed of incorporeal entities: as in the earlier discussion of composing corporeal bodies from incorporeal points, he accepts some sort of limitation to the emergence of properties like corporeality. However, he accepts Aristotle's rejection of the idea that the most basic matter is separable from everything else. At 211,21 Philoponus describes the most fundamental principles as, first, matter – clearly accepting that there is prime matter in Aristotle[45] – and second, the contrary qualities that adhere in it as forms, and third, the elements thus composed.[46]

More needs to be said about the primacy and the role of the four contraries. Philoponus finds grounds in Aristotle for distinguishing between the body which is acting or being affected, and the primary qualities that are the matter of body and which cause change. He sets out to show how other contraries characterizing tangible body can be reduced to the most basic four. Philoponus shows why the universality of touch means that the tangible qualities need to be primary, although sight has primacy over other senses in some respects. The Platonic account is dismissed on the grounds that it would imply that all the elements need to be mixed into everything to explain how everything is visible and tangible. Universal intermixture is dismissed as too inherently problematic.

Alexander's puzzle over why neither heavy and light nor rough and smooth are included by Aristotle is set aside by Alexander's own consideration that only contraries involved in acting and being affected should be considered as fundamental. Philoponus is a little puzzled by Aristotle's list of opposite pairs of properties of tangible bodies, feeling that dense and rare ought to be included. Besides citing Alexander's suggestion that dense and rare reduce to heavy and light, he makes two different attempts to explain the omission, saying once that coarse and fine might capture the meaning of dense and rare, and later suggests that, as the causes of rarity and density, the inclusion of hot and cold might be sufficient.[47]

Philoponus then goes on to consider some problems arising from the relationship of the primary contraries to phenomenal properties. He needs to show that the various phenomenal properties are reducible to the primary contraries, and also that the elements are in fact composed of contraries they may not seem to exhibit. Air in particular draws some discussion, as the claim that it is fundamentally hot and wet is sometimes challenged. Its heat is defended, not by appeal to sense experience, but to those properties that reduce to the primary heat, like lightness. In Aristotle's equation of each element with one contrary that it particularly exemplifies, 'wet' is assigned to air rather than water. Philoponus defends this by arguing that air indeed best fits the definition of wet as that which is easily bounded, rejecting the idea that either the phenomenal quality, or the power to cause the phenomenal quality, should be taken as definitive. The idea that the pair hot and cold are active, wet and dry passive, also needs defence: Philoponus explains that it is not about the relationship of primary qualities to one another, but about their effect on composite bodies.

These chapters lay some of the groundwork for integrating the view that opposites change into one another with the view that qualitative change is ultimately effected by the mixture of the four primary qualities. Philoponus' notion of supervening forms may be aimed at avoiding the perils of a reductive account of qualities or forms while still allowing that the causation of change may be explained by changes in the matter. In much of this he goes beyond Aristotle's views, raising difficulties and proposing new solutions. This is no mere pedestrian commentary, but the work of an independent philosopher.

At the time of his death in Spring 1997, C.J.F. Williams was revising the translation of this section of the commentary, covering chapters 1.6-2.4 of Aristotle's work. As in the companion volume, the aim has been to stay as close as is practical to Williams' draft. Where the notes have been supplemented, these are marked. Because the translation was not yet completely polished, the suggestions and comments of vetters have proved particularly invaluable as have comments from Richard Sorabji and Frans de Haas. Lemmas from Aristotle's text have been supplemented on the basis of Williams' own translation for the Clarendon Press, adapted to accord with differences in Philoponus' text and with Williams' changes in translation style: Oxford University Press kindly gave permission to draw on the Clarendon translation. It is a great pleasure to see Williams' fine work reach publication.[48]

Notes

1. The earlier chapters are published in this series as C.J.F. Williams (trans.), *Philoponus On Aristotle On Coming to Be and Perishing 1.1-5*, London and Ithaca N.Y. 1998.

2. *Aristotle's De Generatione et Corruptione*, Oxford 1982, p. vi. There, Williams left Aristotle's work in its Latin title and translated the subject matter of the work as 'generation' and 'corruption'. Here he uses 'coming to be' and 'perishing', although in other respects his translation of Aristotle often follows his Clarendon version. (Aristotle's work is referred to by its standard abbreviation, *GC*, and Philoponus' work as *in GC*.)

3. An excellent new discussion of this problem is Frans de Haas, 'Mixture in Philoponus. An encounter with a third kind of potentiality', in J.M.M.H. Thijssen and H.A.G. Braakhuis (eds), *The Commentary Tradition on De generatione et corruptione: Ancient, Medieval, and Early Modern*, Turnhout, forthcoming.

4. On the controversy whether the fifth element has matter, see Richard Sorabji, *Matter, Space and Motion*, London and Ithaca N.Y. 1988, pp. 42 n. 69 and 116.

5. In this treatise Philoponus seems to accept an Aristotelian doctrine he later comes to revise, that the heavenly bodies are indestructible and fundamentally different in nature from the sublunary sphere. On his later view, see R. Sorabji, *Time, Creation and the Continuum*, London and Ithaca N.Y. 1983; R. Sorabji (ed.), *Philoponus and the Rejection of Aristotelian Science*, London and Ithaca N.Y. 1987; C. Wildberg (trans.) *Philoponus: Against Aristotle on the Eternity of the World*, London and Ithaca N.Y. 1987 in this series; C. Wildberg, *John Philoponus' Criticism of Aristotle's Theory of Aether*, Berlin 1988. On Philoponus' rejection of the view that the heavens lack tangible qualities, see Wildberg, *John Philoponus' Criticism*, p. 183. On whether causing motion must mean reciprocal effects, see Michael Wolff, *Fallgesetz und Massebegriff*, Berlin 1971, pp. 92ff.

6. Simplicius *in Phys.* 1360,24-1363,24, forthcoming in this series as *Simplicius on Aristotle Physics 8*, translated by Richard McKirahan. For discussion and further references, see Richard Sorabji, *Matter, Space and Motion*, pp. 249-59.

7. This has precedent in Aristotle's remark at *GC* 330b13, where *dêmiourgein* is used to distinguish genuine agency in some pairs of contraries over others. See also Philoponus' use of the term at *in GC* 169,7.

8. Simplicius *in Phys.* 1361,14.

9. There is a large literature on this distinction. Some important sources include J.L. Ackrill, 'Aristotle's distinction between *Energeia* and *Kinesis*', in R. Bambrough (ed.), *New Essays on Plato and Aristotle*, New York 1965, 121-41; Terry Penner, 'Verbs and the identity of actions – a philosophical exercise in the interpretation of Aristotle', in O.P. Wood and G. Pitcher (eds), *Ryle: A Collection of Critical Essays*, Garden City N.Y. 1970, 393-453; A.P.D. Mourelatos, 'Events, processes, and states', in *Syntax and Semantics* 14, 1981, 191-212.

10. The standard Neoplatonist interpretation of Plato's *Timaeus* denies that it should be understood as meaning that the *dêmiourgos* gives the cosmos a temporal beginning. Philoponus, as a Christian, later defends the view that the cosmos could have a beginning in time. See R. Sorabji, *Time, Creation and the Continuum*.

11. This classification is well illustrated in Theophrastus' account of perception: see Han Baltussen, *Theophrastus on Theories of Perception: Argument and Purpose in the De sensibus*, Utrecht 1993.

12. Richard Sorabji pointed out that this argument is in fact one of Aristotle's justifications for an unchanging fifth element, that otherwise all things would be destructible: *Metaph*. 12.5, 1071b5; he argues elsewhere that if the heavens were composed of the element fire into which other elements can transform, then because of the disproportionate quantities, the other elements would have been destroyed (*Meteor*. 1.3, 340a1-3). For Philoponus' reply to this, see Wildberg, *Philoponus: Against Aristotle on the Eternity of the World*, pp. 69-76; Wildberg, *John Philoponus' Criticism*, pp. 194-5.

13. For the view that this model of causation is widely accepted, see A.C. Lloyd, 'The principle that the cause is greater than its effect', *Phronesis* 21, 1976, 146-56; also Barnes' 'synonyny principle': *The Presocratic Philosophers*, London 1979, pp. 88, 119. Alexander P.D. Mourelatos discusses Philoponus' challenge to what he calls, following Lloyd, the transmission model of causality, in 'Aristotle's rationalist account of qualitative interaction', *Phronesis* 29, 1984, 1-16. See Richard Sorabji's Preface to the companion volume.

14. See Richard Sorabji's Preface to the companion volume. At *in GC* 84,3ff., Philoponus discusses a case where something is made hard by a cold thing, taking his departure from Aristotle's remark at *GC* 320b19.

15. Mourelatos discusses these in 'Aristotle's rationalist account of qualitative interaction', p. 7.

16. On Philoponus' view and its later developments, see Sorabji, *Matter, Space and Motion*, ch. 2; De Haas, *John Philoponus' New Definition of Prime Matter*, Leiden 1997.

17. 'Conceptual developments and modes of explanation in later Greek scientific thought,' 61-78 in A.C. Crombie (ed.), *Scientific Change: Historical Studies in the Intellectual, Social and Technical Conditions for Scientific Discovery and Technical Invention, From Antiquity to the Present*, London 1963, p. 73.

18. S. Sambursky, *The Physical World of Late Antiquity*, London 1962, 107-8; E.R. Dodds, *Proclus: The Elements of Theology*, Oxford 1963, pp. 344-5. A key text is Plotinus' *Enn*. 6.4.11.

19. *The Physical World of Late Antiquity*, London 1962, p. 107.

20. G.E.L. Owen, 'Commentary', 93-102 in Crombie (ed.), *Scientific Change*; R.B. Todd, '*Epitedeiotes* in philosophical literature: towards an analysis', *Acta Classica* 15, 1972, 25-35.

21. Todd suggests that the juxtaposition of 'preventing', *kôluma*, and suitability in this passage may be evidence of awareness of the views of Philo the Dialectician: '*Epitedeiotes* in philosophical literature', p. 32. If the notion of suitability is merely a name for Aristotelian possibility, there may be Aristotelian precedents for the idea of external prevention, e.g. *Metaphysics* 9.5, 1048a17. What Philo's discussion and the sceptical response might have highlighted is the fact that causation needs to explain not only why one thing acts on another, but also why causes don't always act, so that features of both agent and patient must appear in the account: Sextus Empiricus *AM* 9, 241-8. On Philo, see David Sedley, 'Diodorus Cronus and Hellenistic philosophy', *Proceedings of the Cambridge Philological Society* n.s. 23 (1977), 74-120.

22. GC 328a33-b5; cf. *GC* 2.2, 329b31. I am grateful to Frans de Haas for a correction on this passage. See Richard Sorabji, *Matter, Space and Motion*, p. 72.

23. For an excellent account of epiphenomenalist accounts of soul in antiquity, see Victor Caston, 'Epiphenomenalisms, ancient and modern', *Philosophical Review* 106, 1997, 309-63. Caston provides an introduction to the modern debates on the topic and discusses the parallels with ancient views.

24. cf. Todd, 'Some concepts in physical theory in John Philoponus' Aristotelian commentaries', p. 167 n. 74.

25. Frans de Haas, *John Philoponus' New Definition of Prime Matter*, esp. pp. 158-62.

26. On this, see Nelly Tsouyopoulos, 'Die Entstehung physikalischer Terminologie aus der neoplatonischen Metaphysik', *Archiv für Begriffsgeschichte* 13, 1969, 7-33; a new paper by Katerina Ierodiakonou discusses this theory in the context of ethics: 'Aspasius on perfect and imperfect virtues', in R.W. Sharples (ed.), *Aspasius*, Berlin 1999.

27. Simplicius *in Cat.* 288,18-30; cf. 284,14ff. A new translation by Barrie Fleet, *Simplicius: On Aristotle Categories 6-8*, is forthcoming in this series.

28. Katerina Ierodiakonou discusses possible Stoic influences in her paper, 'Aspasius on perfect and imperfect virtues'. Todd expresses reservations about the applicability of the *Categories* controversy to Philoponus' usage, as the discussion at *in GC* makes no reference to participation of a qualified thing in an unenmattered quality: he points to another Aristotelian source for the notion of a *epitasis* and *anesis* in a discussion of motion in *Cael.* 288a19: 'Some concepts in physical theory in John Philoponus' Aristotelian commentaries', *Archiv für Begriffsgeschichte* 24, 1980, 151-70, pp. 168-9. At *in Cat*, 159,21-4, Philoponus talks of qualities unintensified or relaxed in themselves, but intensified or relaxed in the bodily substrate.

29. De Haas points out that this idea becomes explicit in *in GC* 2.7: see his 'Mixture in Philoponus. An encounter with a third kind of potentiality'.

30. The terminology of *platos* is absent here, but the doctrinal similarities are pointed out by De Haas, *John Philoponus' New Definition of Prime Matter*, pp. 156ff. cf. Wolff, *Fallgesetz und Massebegriff*, pp. 138-48.

31. *John Philoponus' New Definition of Prime Matter*, pp. 161-2. One source for this doctrine may be the Stoic view that while there are stages of cosmic conflagration when all elements turn into fire, it is nonetheless true that the size of the cosmos remains finite even during maximum expansion. So, the quantity of fire that can be produced from the entire mass of each element must be fixed.

32. *Meteor.* 1.3, 340a12-13.

33. *in Phys.* 97,24-7; 103,22-5.

34. cf. Michael Wolff, *Fallgesetz und Massebegriff*, p. 132. The term *atomos* is applied to flesh in the same way it is to a composite or to a mixture, to mean that it is point-like or individual. See *in GC* 1.8, 170,13-14 for the contrast between qualities that are *atomos* and those that occur in a range.

35. *Contra Proclum* 434,9ff. See De Haas, *John Philoponus' New Definition of Prime Matter*, 161-4, for discussion.

36. *in Phys.* 98,21-6.

37. A.P.D. Mourelatos, 'Aristotle's rationalist account of qualitative interaction', *Phronesis* 29, 1984, 1-16.

38. Although its authenticity was questioned, recent scholars have accepted this chapter, as the Neoplatonists clearly do, as being by Aristotle. For an overview of the recent scholarship, see the introduction to Eric Lewis (trans.), *Alexander of Aphrodisias: on Aristotle Meteorology 4*, London and Ithaca N.Y. 1996, p. 15, in this series.

39. Classic sources on this controversy include Friedrich Solmsen, 'Greek philosophy and the discovery of the nerves', in *Kleine Schriften*, Hildesheim 1968; Jaap Mansfeld, 'Alcmaeon: "Physikos" or Physician? With some remarks on Calcidius' "On Vision" compared to Galen, *Plac. Hipp. Plat.* VII', 26-38 in Mansfeld and de Rijk (eds) *Kephalaion: Studies in Greek Philosophy and its Continuation Offered to Professor C.J. de Vogel*, Assen 1975; David Hahm, 'Early Hellenistic theories of vision and the perception of color', 60-95 in Machamer and Turnbull (eds) *Studies in Perception: Interrelations in the History of Philosophy and Science*, Columbus 1978.

40. See notes 176-9 to the translation below.

41. Simplicius actually finds six different senses of *dunamis* in Aristotle *in Cat.* 225,1: see Barrie Fleet (trans.) *Simplicius: On Aristotle Categories 6-8*, forthcoming in this series.

42. On this problem see De Haas, 'Mixture in Philoponus. An encounter with a third kind of potentiality', which discusses whether numerically the same ingredients are recovered from a mixture. Richard Sorabji discusses a difficulty for Aristotle over whether numerically the same ingredient is recovered and also a distinction between the loss of an ingredient and the loss of its power, at *Matter, Space and Motion*, pp. 67-9.

43. See 'Mixture in Philoponus. An encounter with a third kind of potentiality'. De Haas notes that Alexander is a plausible source for the notion that the components of true mixtures cannot be the same as they would be in cases of mere juxtaposition. The definitive mark of the difference is that some addition is needed to resurrect the pure capacity.

44. 'Mixture in Philoponus. An encounter with a third kind of potentiality'.

45. On this debate see Williams' Appendix to the Clarendon edition, *Aristotle's De Generatione et Corruptione*, translated with notes by C.J.F. Williams, Oxford 1982, and the companion volume of Philoponus' commentary, *Philoponus On Aristotle On Coming-to-Be and Perishing 1.1-5.*

46. For the role of the three-dimensional in this, see Frans de Haas, *John Philoponus' New Definition of Prime Matter*, esp. pp. 137-53; also *in GC* 1.5 in the companion volume.

47. Much of Alexander's lost commentary on *GC* 2.2-6 has been tracked down in two Arabic versions by Emma Gannagé: this shows, p. 68, that the last suggestion made by Philoponus is in fact raised by Alexander. See Richard Sorabji's Preface, above.

48. I am very grateful to Richard Sorabji and Frans de Haas for helpful suggestions and comments on earlier versions of the Introduction.

Textual Emendations

134,6 Reading *antipaskhein* for *antipaskhei*
140,24 Reading *to oligon pur hupo tou pollou* rather than *to polu pur hupo tou oligou*
148,11 Reading *tôi hudati* instead of *tôi puri*
148,32 Following the Aldine editor in omitting the words *ean ara*
151,12 Following Vitelli's addition of *en* here
152,13 Following the Aldine editor in adding *tên*
152,14 Following Vitelli's suggestion to add *oun* here
154,21 Following Vitelli in understanding *hoi gar peri Parmenidên*
156,5 Reading *autôn* rather than *autois* here at Furley's suggestion
158,1 Reading *mêden* for *mê dein* at Furley's suggestion
164,26 Following the Aldine editor's insertion of *hapsasthai. ei men oun en tôi kenôi khôrêsai, ouk aisthêsetai. hên* (sic) *de tês atomou*
165,32 Reading *elegon* for *elegen*
166,19 Following the Aldine editor in inserting *pros to tou heterou kenon*
168,10 Following the Aldine editor in omitting *oude*
171,7 Following Vitelli in deleting *to kuriôs on to auto to de pantelôs êmeiphthai*
177,18 Following the Aldine editor's addition of *ara*
197,27 Reading *mignumenê* with two MSS and the Aldine editor, rather than *mignutai*
201,7 Following Joachim in omitting *ê* from *pampan ê êrema*
225,8 Reading *hen ti tôn tessarôn stoikheion poiountes*
231,7 Taking *leptotêtês* to be a misprint for *leptotêtos*

Philoponus
On Aristotle
On Coming-to-Be and Perishing 1.6-2.4

Translation

Lecture Notes taken by John the Philologist of Alexandria from the Course given by Ammonius the Son of Hermeias with some Observations of his own on the First Book of Aristotle's *On Coming-to-Be and Perishing*

<1.6. CONTACT>

> **322b1** And since we must first speak about the matter and the 123,28
> so-called elements, whether or not they really are such, and
> whether each is eternal or they in some way come to be [and, if 30
> they do come to be, whether they all come to be from one another
> in the same manner or whether some one of them is primary ...]

He has refuted the false opinions of the ancient <philosophers> about
coming-to-be, has given a general account about coming-to-be and
distinguished it from alteration and spatial movement, has said what
growth is and what it is that grows, and by what it grows and what
is the manner in which growing things grow.[1] Next it would follow 124,1
appropriately to speak of the coming-to-be of the particular <kinds
of> things that come to be, e.g. meteorological phenomena, plants,
animals, minerals.

But a person who intends to speak about the coming-to-be of
particular <kinds of> things ought first to discuss the matter proxi-
mate to each <kind of> coming-to-be, e.g. what the matter is which 5
underlies the things that exist in the atmosphere, what is that which
underlies the minerals, and the rest in the same way. For just as the
account of natural things in general was prefaced by an account of
matter *simpliciter*, i.e. that which is the common substratum to all
natural objects, which is the topic of his teaching in the *Physics*, so
the <account> of particular <kinds of> coming-to-be and the matter
underlying each of them needs to be prefaced by the <account> of the 10
matter that belongs in general to the things that come to be *simplici-
ter*, i.e. the elements.[2] This is why it was necessary to set out first the
<account> of the elements. For someone who does not know what the
elements are, and how many there are of them, and what they are
like cannot discover the special matter proper to each <type of thing
that comes to be>; for it is these, after prime matter, which are the

common substratum for everything, and they are the corporeal matter common to all the things that come to be.

This is why he plans to examine them first, whether or not these four that are constantly mentioned are the elements of the things that come to be, whether they are capable of coming-to-be or incapable and unchangeable, as Empedocles thought, and if they are capable of coming to be, whether they come to be from each other or from something else, and if from each other, whether each of them changes equally into each of the others or whether there is one of them which is first, from which all the others change, as it were matter underlying the rest, as Thales thought was the case with water, Anaximenes with air and Heraclitus with fire. And he will enquire whether they are four in number or more, and will show that they are four only. These questions delimit the subject matter of the treatise we are studying. And so finally in the treatises which follow he produces accounts of the particular <kinds of> things which come to be, in the *Meteorology* an account of meteorological phenomena, in *On the Generation of Animals* the coming-to-be of animals, and in a similar manner of the other things. Already indeed in the treatise *On the Heavens* he has investigated the elements and has shown that the discussion of the elements comes first in order, and also that they are neither incapable of coming-to-be nor imperishable, and that they do not come to be from something else which underlies them, and that they are not unlimited but limited. In that work, however, he omits any discussion of the question of their number or of the manner in which they come to be from one another, intending to discuss these matters in the book we have in front of us.

The object therefore is to demonstrate the nature and the number of the elements, and that their coming-to-be is from one another and that there is just one way in which they come to be from one another. But there are one or two topics preliminary to the discussion of the elements, and if they are not sorted out it is impossible for the discussion of the elements to proceed conveniently. These are what he proposes to speak of. There are three of them: What is mixture?, What is acting and being affected? and What is contact?

He draws also on his predecessors to provide reasons for holding that it is necessary for someone who is to discuss the elements to give an account of these matters. For all of them, he says, both those who said that the elements were many and those who said that they were one, appealed to acting and being affected,[3] aggregating and segregating. For those who said that there was one element and that the other things came to be by aggregation and segregation of this appealed to acting and being affected; for being aggregated and segregated is being affected in a way. And if there is something that is affected, there must necessarily be something that acts. And the same is true of those who say <the elements> are many, both of those who say they are changeable and of those who say they are unchangeable. Those who say they are changeable obviously appeal to acting and being affected, since it is by acting upon each other and being affected by each other that they produce change in one another. Empedocles too, who says they are unchangeable, appeals to acting and being affected; for even if they are not affected by change in respect of their substance, nevertheless they undergo aggregation and segregation, which are brought about by strife and love.[4]

So an account of acting and being affected is essential for someone who is going to discuss the elements. For how shall we get to know in general whether they are impassive or not if this very issue, of what in general acting and being affected are, is something about which we are ignorant? But if it is necessary to speak of acting and being affected, obviously, he says, it is necessary also to discuss mixing. For things which act upon and are affected must first have been mixed with each other, since they would not act upon or be affected by one another while they were bodies standing at a distance from each other.[5] Clearly too those who say that coming-to-be occurs through aggregation and segregation are bound to employ <the notion of> mixing; for aggregation is a sort of mixing.

So if a discussion of mixing is needed, without doubt contact must <be discussed> too, since things that are mixed with each other must first touch each other; for how should things which have not touched be mixed? So for an account of elements an account of acting and being affected is necessary, and for this an account of mixing (for the things which act and are affected must have been mixed with each other), and for this an account of contact; for how should things that have not touched be mixed with each other? So we have to have an account of contact and of mixing and of acting and being affected. And in any case, he says, there is a need to produce a definitive account of these things, because the philosophers of nature who preceded us produced no account of them, <no explanation of> what each of them might mean, but left them in that way undefined. So he begins by giving an account of these, and first of contact.

126,1 'And since we must first speak about the matter and the so-called elements.' Since he has said 'about the matter', he adds 'and the so-called elements' to prevent our supposing that he is speaking about prime matter, that which is incorporeal, which he has already examined in the *Physics*.[6] It is as if he had said 'about the matter which is the so-called elements'. His reason for saying 'so-called' is perhaps that he has not yet shown that these are the elements. 'Whether or not they really are': i.e. whether they are the elements of the things that come to be; for the <question is> not whether they are *simpliciter*, since this is evident. 'Or they in some way come to be': He adds 'in some way' as an alternative to 'in a certain manner', because they do not come to be as wholes, but as parts.[7]

> **322b5** ... before this we must speak of things that are now talked about in a confused way.

Here <we have> the apodosis for the clause <beginning> 'And since': 'And since we must first speak about the matter and the so-called elements, before this we must speak of things that are now talked about in a confused way.' And he speaks of contact and mixing and acting and being affected as contributing to his theory of the elements and as not having met with a definition and explication from his predecessors.

> **322b6** All <philosophers>, both those who make the elements come to be and those who make the things derived from the elements come to be, make use of aggregation and segregation, acting and being affected.

These remarks enable him to establish that the discussion of the things mentioned is necessary, given that all those who treat of coming-to-be say that it is by aggregation and segregation and by acting and being affected that things come to be. But one is justified in wondering what he has in mind by saying that they all make use of segregation and aggregation, acting and being affected. Whereas it is evident that they all used segregation and aggregation, it is not the same with acting and being affected, but only those who make things come to be by the arrangement of the substratum,[8] since those who believed that the elements are unchangeable did not make use of acting and being affected but only of aggregation and segregation.

So either he is saying that some of them make use of segregation and aggregation, others of acting and being affected and thus by synecdoche[9] says that they all use both, but separately (for those who only make things compounded of the elements come to be made use of aggregation and segregation, and these were the schools of Empedocles, Democritus and Anaxagoras, because each of them regarded his own primary kinds of thing[10] as impassive; those on the other hand who allow the elements also to come to be <made use of> acting and being affected; and these were the people who posited some one thing out of which they made the rest come to be, and this one <they took> to be either one of the four or something else over and above these) – so either this is the sense of his remark about their all making use of both aggregation and segregation and acting and being affected, or he is saying that all of them used both <sorts of mechanism>. For he makes clear that this is what he is saying by his remark a little later:

'For those who posit several elements make them come to be by acting and being affected, and equally those who <make them come to be> from one <element> cannot avoid speaking of action.'

So here we are to understand acting and being affected, not in terms of the substratum being arranged and changed, but simply in terms of the efficient or motive cause itself being actualised.[11] For it is necessary in each instance for there to be a cause of aggregation and segregation. For Empedocles, on the one hand, strife and love are the causes, the latter of aggregation and the former of segregation, and to aggregate and segregate is to act in some way, so that to be segregated or aggregated is to be affected. For Anaxagoras, on the other hand, the producer of aggregation and segregation is Intelligence. And for that matter according to Democritus too shape, arrangement and position[12] would be efficient causes, and in addition chance and accident.

322b8 [Now, aggregation is mixing;] but what we mean by 'mixing' has not been clearly determined.

<Is mixing produced> by the elements being cut up small and placed alongside each other and escaping detection by the senses by their smallness, in the way that we might mix fine-ground meal from wheat and barley (for the whole would appear to be one type <of stuff>, the senses being deceived by the smallness of scale)? Or would it consist in their all being present in each part and the whole of the ingredients passing through one another as a whole? And do the forms of the elements perish in the mixing, or are they preserved? And if preserved, is <this the case with> all <of them>, or <only with> some?

322b9 What is more, it is impossible for there to be alteration 128,1
or segregation and aggregation, if there is nothing which acts nor anything which is affected.

For unless there were something which alters there would not be anything which is altered. But how would it be possible for something to be altered unless the thing that was altered were affected in some way by the thing which altered it?

322b12 [For those who posit several elements make them come to be by acting and being affected,[13]] and equally those <who make them come to be> from one <element> cannot avoid speaking of action. And Diogenes is right to say this, that if it were not the case that everything is from a single thing, there would not be any acting upon or being affected by one another,

> [e.g. what is hot being cooled, and vice versa – for heat and cold do not change into each other. What changes is clearly the substratum; so objects between which there is action and passion necessarily have a single underlying nature].

For if they were using the alteration of the one to explain the coming-to-be of the others, depending on it being condensed or rarefied, it is obvious that they would have been using acting and being affected. For in alteration the thing that produces the alteration acts and the thing that is altered is affected. And he supports the claim that those who suppose that the element is one are forced to speak of acting and being affected by calling as a witness Diogenes of Apollonia, who says rightly that if the underlying thing were not one <and the same> for everything, they could not act upon and be affected by one another, on the grounds that their having one thing as substratum follows necessarily on their acting upon and being affected by one another. He shows that if the underlying thing were not one they would not be able to act upon and be affected by one another, because Diogenes says that it is not heat which changes into cold, or vice versa (for it is impossible for the contrary to come to be what its contrary is), but that which underlies them, being common to <both> the contraries and affected by one of them changes into each <state> according to the action of the agent.

Plato also says this in the *Phaedo*. For having said previously[14] that contraries come to be from one another, the white from the black and the black from the white, the living from the dead and vice versa, he goes on to say[15] that it is never possible for the contrary to change into its contrary; for whiteness could not become blackness nor coldness heat. Because someone was puzzled about this, about how Socrates avoids being shown to contradict himself, <Plato> says that earlier he had spoken about contraries coming-to-be from each other – not the things participated in, I mean the contrarieties themselves, but the things which participate in them (for the white body comes to be from the black body) – but now, he says, he is saying that the things which are participated in do not come to be from one another.[16]

But it would be more accurate to speak of the <contraries> as
129,1 <coming to be> after one another;[17] for the blackness comes to be after the whiteness, the quality itself perishing while the substratum remains.

> **322b19** But it is not true to say that everything is of this kind, but only those things between which there is interaction.

He remarks that it is not true to say that all the things that are are of this sort, namely, such as to have one and the same matter underlying them. Or this <is what he is saying>: it is not true to say that there is one matter for all the things that act and are affected. For the body whose motion is circular itself acts and something is affected by it, namely all sublunary <bodies>, but it is not the case that that which acts and that which is affected are of one and the same matter.[18] But this, he says, is how we must speak: the things which have a common matter are those things which are mutually affected by each other. For this is the reason for their being *mutually* affected by each other, namely, the fact that one matter is subjected to them, one which is receptive of opposites. This indeed is why the heavenly <bodies> act upon the occupants of this region without being affected in return by them, namely, the fact that there is no one matter that underlies them.[19]

> **322b21** Furthermore, if consideration has to be given to action and being affected and to mixing, it must necessarily also be given to contact.

He has shown that an account of acting and being affected and of mixing is needed: from this he shows that an account of contact is needed too.

> **322b22** For neither is acting and being affected possible in the strict sense for things which cannot be in contact with one another, nor can things that have not touched in some way be mixed. [So we must get clear about these three things: what contact is, what mixing is, and what action is.]

It is impossible, he says, for things that in the strict sense act and are affected not to be in contact with one another. For it is by touching the patient either immediately, like the fire <touching> the fuel, or mediately, like the fire touching the water through the pot as an intermediary, that the agent acts upon the patient. He adds 'in the strict sense' because we are said to be affected by slander, but slander would not be said to act in the strict sense of the word, because it does not touch either, given that acting is changing natural bodies in respect of quality. But if the slander is also said to touch the person slandered, this too is not said in the strict sense of the word, given that those things are said to touch whose extremities are together. So in the sense in which slander might be said to touch us, in that
sense <it might> also <be said> to act upon us; neither <is so 130,1
described> in the strict sense of the words, as we shall learn in the following passage.

'Nor can things that have not touched in some way be mixed.' The structure of the sentence is an example of hyperbaton,[20] for the sequence of thought is like this: nor can things be mixed unless they have first touched in some way. He adds 'in some way' to signify things that have touched in the strict sense, since slander too is said to touch <people>, but not in the strict sense <of 'touch'>.

> **322b26** Let us make the following start: it is necessary for those of the things which are which are involved in mixing to be capable of contact with one another, [and the same holds for anything which properly speaking acts on, or is affected by, another. So the first thing to discuss is contact.]

He has shown that, in order to establish the fact that the four elements are the matter of things that come to be and perish, it is necessary <to give> an account of contact, mixing and acting and being affected. He intends to begin by discussing these, and first, touch. He has already shown that the account of touch has precedence over the other two: now once again he takes up the same <theme> before <giving> his account of contact. Those of the things which are whose nature is to be mixed must first be in contact with one another; for things which were in no way touching but separated <from each other> would not mix. So the account of contact must precede that of mixing.

And the same holds good for acting and being affected. For if there is one of the things that are whose nature is to act and another whose nature is to be affected <by it> in the strict sense of the words, these must first be in contact with one another. This he has proved adequately in what has gone before. He says 'in the strict sense' because of slander, since this is said to have an effect on the mind, whereas it does not touch the mind. For it does not act in the strict sense because it is not in the strict sense that it is said to touch; so it is said to touch in the same sense as that in which it is said to act. The same could be said about likenesses and fodder and the object of desire in general; for these too do not act in the strict sense. But in so far as they do act, to this extent they first touch. For first they touch the imagination and make an impression on it; then they change it by producing an alteration <in it>. And it is impossible for anything to be affected by the object of desire if it has not been in contact with it by receiving an impression on its imagination. So he is obliged first to undertake an account of contact.

Many other words are used in multiple ways; and of those which are used in many ways some are used homonymously[21] (e.g. 'Ajax' both of the son of Telamon and of the son of Oïleus, and 'dog' both of the land animal and of the marine animal), which have an equal share
131,1 in the name, others have a common origin or a common focus.[22] In their case one is used in a stricter sense and the other in a less strict

(e.g. the case of 'being', which primarily and strictly applies to substance and secondly and thirdly to the other categories; similarly the real human being is called an 'animal', but the painted one[23] is called an animal by extension from the real one and is not called 'animal' in the strict sense; similarly knowledge of music[24] and the musical woman, <called so> from it, and the musical lyre[25] – the word 'musical' is manifestly predicated primarily and in the strict sense of knowledge of music). So in this way 'contact' too is one of the words which are used in many senses, <though> with a common origin and a common focus, and one of these senses of 'contact' is the strictest, and the less strict sense is derived from it.

So he tells us what contact is in the strict sense and what the things which are in contact in the strict sense, namely, that they are those which have position, i.e. are situated somewhere. But mathematical objects too belong to those objects which have position, so these too would count as being in contact with one another. For we say that the line touches the line or the surface, e.g. a circle or some other thing. But the mathematical body, since it has extremities, would also touch either a mathematical body or a line. But even if mathematical bodies were said to touch, this would not be in the strict sense, but <only> in the way that they also have being. They have this in thought: in this way then they also have position and touch one another. And this is true of them in virtue of the things which have position and touch in the strict sense of the words, I mean natural bodies from which the mathematicals are abstracted.

He constructs his account of how the things that have position are also in contact with one another from his definition of things in contact, given in *Physics* 5.[26] There he says that being in contact is this: things' having their extremities together. But if this is what being in contact is, those things would be in contact which, being discontinuous objects having size – i.e. not unified but circumscribed by their own surfaces, with their own positions – have their extremities together. For generally things which do not have size cannot have extremities either, and furthermore things with size that are not separated but joined together cannot have different extremities.

And these things also have position, and this is the same as being in a place. Places are most properly differentiated by two things: above and below; and it is the things that have either heaviness or lightness that are above and below. These are things which are capable of acting upon and being affected by each other. So it can be concluded from all this that those things touch one another which are capable of acting upon or being affected by one another. Those things then are in contact in the strict sense of the word whose nature is to act upon and be affected by each other. As we continue with our examination of the text we shall discuss problems which arise in connection with this account.

132,1 **322b31** [Just as almost all other words are used in a variety of senses,] some of them equivocally, others with some of their senses dependent on other, prior senses, [so it is with ‘contact’.]

It should be noted that he does not consider <expressions> with a common origin and a common focus to be homonymous, but <to belong among> those used in many ways but not homonymous. For <the class of expressions> used in many ways extends further than <that of> the homonymous. And if in the *Categories*[27] he classifies <expressions> with a common origin and a common focus amongst the homonymous, we say that in the sketchy survey he conducts there he calls expressions used in many ways homonymous, so that the class <of expressions> he is determining there is of those which are used in many ways. Those then that have a common origin and a common focus are midway between the homonymous and the synonymous.[28]

322b32 But <contact> in the strict sense belongs to things which have position, and position to things which also have place.

He identifies the sort of things which, in the strict sense of the words, are in contact as those which have position, and these are those that are in a place; for this is what it is to have position, namely, to be in a place. That is all for the moment that he says about things that are in contact, but he is going later to add that they are bound also to be things that are capable of acting and capable of being affected.

323a1 For even the mathematical entities, to the extent that contact is attributed to them, must also have place, whether each exists as a separate object, or in some other way.

Since he has said that things that touch also have position and are in a place, the thought occurs to him ‘Why then, since mathematical magnitudes touch one another, e.g. the line and the circle,[29] does it follow that they also have position and are in a place?’ This is why he says that these too analogously to <cases of touch> have position and place. For in this way natural objects differ from mathematical objects, by the former admitting affections, I mean heaviness and lightness and the others, whilst the latter have being only in thought without affections and in respect only of extension. But being in a place is a property of bodies, not in so far as they have affections, but in so far as they have extension. And the mathematical <body> also has extension; so this too is in a place. But if the mathematical are 133,1 separable from the physical <bodies> and subsist *per se*, they also have a place *per se* and could be said to be in contact in the strict sense of the words. If, however, they are not separable from the physical bodies, but have their being in these <bodies>, which is indeed the truth, it is quite clear that being in a place and being in contact with each other will belong to them in the same way in which being does. So if they have their being in those they will also, no doubt, have <the property of> being in a place and being in contact with one another because of the <physical bodies>.

> **323a3** If <to be in contact>, as was defined earlier, is [to have extremities together, those things will be in contact with each other which are discontinuous objects having size and position and which have their extremities together.]

This is how being in contact is defined in Book 5 of the *Physics*,[30] as having the extremities together. The words 'discontinuous objects having size' are well chosen, because of the parts of the continuous; for the parts of the continuous, not being discontinuous objects having size, but existing potentially in the whole, are not said to be in contact either, but precisely this: to be continuous and one.

> **323a6** Since whatever has a position also has a place and the primary distinguishing features of place are 'above' and 'below' [and the other opposites of this sort ...]

By this means he intends to sort out what are the things which are in contact in the strict sense of the words, and which are in contact physically (for the topic of the discussion is the things which are in contact physically), namely, he says, those which have position and are in a place and act upon and are affected by each other; for the things which merely touch[31] one another are in contact mathematically not physically. An objection could be made to the effect that his conclusion, that things that have position in all cases also have weight or lightness and act and are affected, is at fault. For the spheres that are inside that of the fixed stars have position and are in a place and in contact with one another, and yet do not act upon and are not affected by one another, according to Aristotle's own opinion. How, then, does he come to say that the things that are in contact must in all cases be capable of acting and being affected? For either the spheres are not in contact with one another, but are either continuous or at a distance from each other, which is unreasonable, or else, if they *are* in contact with one another, there is no necessity for things that are in contact to act upon or be affected by each other.

134,1 So either he is taking the celestial body[32] to be a single thing, which neither has position nor is in a place (for nothing external contains it, and in this sense it is not in a place, according to Aristotle's stated definition of place, namely, that place is 'the boundary of the container in so far as it contains the thing contained'), or else the present discussion is not concerned with this at all but with things which are in contact within <the sphere of> coming-to-be and perishing, which both share a common matter <enabling them to> act and be affected in response and are in all cases either above or below. And if the definition he gives in *Physics* 5, that things which are in contact are those whose extremities are together, is a good one, there is no need for all the things that are in contact to be capable of acting upon each other and capable of being affected. For the spheres that are within the sphere of the fixed stars are in contact with each other, but they are nevertheless incapable of being affected. If, however, it *is* necessary for things that are in contact to be mutually affected by each other, the definition of being in contact given in the *Physics* is not correct, since it includes more things.

His description of above and below as the primary differentiating features of place is correct. For these are both the primary and the only differentiating features of place, by which the physical bodies are indeed distinguished from each other. For the other contrasting features, in front and behind, left and right, are not strictly speaking contrasting features in respect of place, but depend for their existence on the relation <of the objects in question> to us.[33] So there is one contrast in terms of place, above and below, as is demonstrated in *On the Heavens*. For every <instance of> motion in a straight line is either from the centre towards the circumference or from the circumference towards the centre, and we call one of these 'upward' and the other 'downward'. The other opposites, as I have said, depend for their existence on relation; for that which is right for one person is left for another, and that which is in front for one is behind for another. Moreover, these movements, if they are not going around in a circle but are in reality proceeding in the same direction *ad infinitum* to the right, outstrip the universe.[34] So their movement too is towards the circumference; but this is what <movement> upwards was. And the same applies in the other cases.

323a8 [... all things that are in contact with each other] will possess heaviness or lightness, either both or just one.

The extreme <elements>, earth and fire, possess just one (for the one is light and nothing else, and the other is heavy and nothing else), whereas the intermediate <elements>, air and water, possess both lightness and heaviness, as is demonstrated in *On the Heavens*;[35] for they are heavier than the light things and in this sense have heaviness, and are lighter than the heavy things and in this sense have lightness. So the discussion is not concerned with the body whose motion is cyclical, since this has neither lightness not heaviness.

But if what is said is to be understood as applying to the body whose motion is circular as well, in order to make the account more general, then, in Alexander's opinion, 'either both or just one' must be taken to refer, not to lightness and heaviness, but to the things that are in contact themselves.[36] For it is in all cases necessary that there should be two things that touch, and these touching things, either both or just one of them, necessarily have either heaviness or lightness. <Alexander> also says that 'by one another' where <Aristotle> says 'being capable of changing and being changed by one another' is to be understood as an alternative to 'one by another', as in the *Categories*[37] <Aristotle> uses the phrase 'genera under one another' <to mean>, not each genus being <subordinate to> the other, but one to another.

> **323a9** Now things like this are capable of being affected and of acting; [so it is evident that the things whose nature is to be in contact with one another are those discontinuous objects having size whose extremities are together and which are capable of changing and being changed by one another.]

This is not the conclusion of a syllogism (for it does not follow that because things in contact have heaviness and lightness they are capable of acting upon and of being affected by one another). But because of its very obviousness he concludes this: that things distinguished by <their being in> these places are 'capable of acting and of being affected' (instead of 'capable of changing and being changed,' for he infers this immediately). They are seen more clearly being changed by each other in respect of change of place;[38] for things that are heavy and light are thrust aside one by another as they rush towards their own locations.[39] Or else he has shown potentially that these things are capable of acting and of being affected by calling the places in which they are 'contrary'. For things that are in contrary places, having contrary powers, must needs be capable of acting upon and being affected by each other.[40] And if we understand the passage in this way, evidently the discussion is not about the body whose motion is cyclical, but about things that share a common matter.

But it is possible, as I have said, to extend the account to <cover> the body whose motion is cyclical as well by taking 'by each other' to mean 'one by another'. However, if we take the celestial body to be one thing, there will be nothing opposed to it, whereas if we consider the inner spheres as being in contact with one another, how is it still possible to maintain that they are capable of acting upon and being affected by each other, or that in general they possess heaviness or lightness? So being in contact in a more general sense is, as he says in the *Physics*,[41] having the extremities together (and in this sense celestial bodies are in contact with things here[42] and with one another, but a more special <meaning> for it, one more relevant to natural science,[43] <is> having the extremities together <combined> with having the power to act upon and be affected by each other, as he says here.

> **323a12** Since not everything that changes something changes it in the same way, but in some cases the changer in changing is also itself necessarily changed, while in other cases it is unchanged, clearly the same distinctions will have to be made with regard to the thing that acts; [for the changer is said to do something, i.e. act, and conversely the agent is said to change things. All the same, there is a difference between them and they ought to be distinguished; for not every changer is capable of acting – if, that is, we oppose 'that which acts' to 'that which is affected', a term which applies to those whose change is affection, affection being restricted to that in respect of which a thing alters, such as whiteness and heat – but changing extends wider than acting does.]

He has explained what the things are that are in contact in the strict sense, namely, discontinuous objects having size and being in a place and having their extremities together – things which are capable of changing and being changed by each other. He now follows up the other things, those which are themselves also said to be in contact, but not in the strict sense like the former. Since he has demonstrated in the *Physics*[44] that not everything that causes change does so while being changed, he now takes this and uses it to show that not all the things that touch touch things which <in turn> touch them, but that in their case the word 'touch' is not strictly applicable, but <only> in the wider sense. Taking as a premiss the proposition that not everything that causes change does so while being changed, he assumes that it follows from this that not everything that acts is affected, and he argues for this on the basis also of what some people think about acting and causing change, a view which he then refutes. For some people think that causing change and acting are the same thing, and that these propositions are convertible: 'If something acts, this also in every case causes change' and conversely, 'If something causes change, this also in every case acts'.

So if not everything that causes change does so while itself being changed, and causing change is acting, then neither is everything that acts[45] necessarily in all cases affected, but even if this proposition 'Not everything that acts is affected' is true, it is nevertheless false to suppose that causing change and acting are the same thing. For these differ from each other in terms of being more universal and more particular: causing change covers more things than acting. So if something acts, this in every case causes change, but it is not also true that if something causes change, this also necessarily acts. And he argues for this from its opposite, I mean being affected; for if we knew what being affected was, we should also recognise what acting is. Being affected therefore is changing[46] in respect of quality, e.g. becoming white, becoming hot, and generally being altered in respect of quality. So if this is what being affected is, this then is what acting is: changing <something> in respect of quality. So of the things which cause change, those that bring the things they change to an affection clearly also act, but those which do not bring the things they change to an affection cause <them> to change but do not act.

The things that cause change in respect of place do not bring the things they change to an affection:[47] the celestial body changes <things> in respect of place, but does not alter <them>. Causing change, then, covers more than acting. And here you have the explanation of why Aristotle does not say openly that God is capable of acting upon the world.[48] For if acting is changing <things> in respect 137,1
of quality, and God is creator of the substance of the universe, obviously he would not, strictly speaking, be said to act upon the world, but rather to create and produce it.[49]

> **323a20** This, then, is clear: that there is a sense in which the things which cause change will touch the things they change and a sense in which they will not.

Having shown that causing change covers more than acting, he proceeds to his account of things in contact, which was his reason for mentioning things that cause change and things that change. He says then: since, of things that cause change, some cause change while themselves being changed, while others do so without being changed,[50] this shows that some of the things that cause change also touch the things they change and some do not. For in the case of the things which are changed and are themselves in turn causes of change 'touch' is predicated of them in the strict sense of the word. But in the case of those which cause change and are not in turn changed themselves, 'touch' is not predicated in the strict sense. For neither the image nor the slander 'touches' the soul in the strict sense of the word, since they are not in turn affected by it. For in these cases one would not say either that their extremities are together, which has to be present in all cases for things to be touched in the strict sense of the word.

> **323a22** But the definition of 'being in contact' in general is applicable to things which have a position and are capable the one of moving and the other of being moved, whereas that of 'mutual contact' is applicable to things capable of moving and being moved in which there is acting and being affected.

Having said that not all the things that cause change are also in contact with the things they change, he now gives definitions and says which, amongst the things which cause change, are the ones which are in contact; and he lists three things as being signified by 'being in contact', two more general, one more specific and strictly so called. What he says are in contact in the <more> general sense are things that have position, that is, those that are in contact by way of *touching*,[51] and in this sense mathematical objects too are said to be in contact with one another, as well as celestial objects both with each other and with things here <in the sublunary sphere>; for he himself says that both the line and the surface have position, and if things which have position also had a place, both lines and celestial bodies would be in a place. But what would be the place of a line?

However, as Alexander says, 'position' would be meant by him in the <more> general sense of the order which things which are capable of being changed have one to another, and the simultaneous support and permanence of the cluster, as he says in the *Categories*.[52] And it seems that in the way that these things have being, they also have place, as he says himself: 'for even the mathematical entities, to the extent that contact is attributed to them, must also have place'. So
138,1 this is the first and most general meaning of 'things in contact', as he defined it also in *Physics* 5, namely, things in contact are those whose extremities are together, whereas the second meaning involves the one causing change and the other being changed: for he says, 'and are capable the one of being moved and the other of moving'. According to this <definition> mathematical objects are not in contact with one another, nor those things where it is not the case that the one is capable of acting and the other of being affected, but celestial <bodies> do touch those here,[53] since the one acts and the other is affected.

The third <meaning>, which is most specific and in his view <what the word means> in its strictest sense, as in the case of those things which are physically in contact, he says that of the things which cause change and are changed, those are in contact amongst which there is a mutual <process of> being changed by and being affected by one another. For this is what he means by the words 'whereas that of "mutual contact" is applicable to things capable of moving and being moved in which there is acting and being affected'. The word 'mutual' represents the idea of <being affected> by each other, and amongst these he no longer counts celestial <bodies>. So either these are counted as three, or still as one, though after initially, without further distinction, saying that things in contact are those which have position and of which one acts and the other is affected, he proceeds later to make nicer distinctions by adding that these must be mutually caused to change and be affected by each other.

> **323a25** The more usual case, indeed, is where a thing touches something which in turn touches it; for practically all the things we meet cause change while being changed.

Having discussed the things that are in contact in the strict sense, he turns to those that are said to be in contact in the more general sense. He says that for the most part everything that touches touches something which touches *it* – the case of touching in the strict sense; for if things that are in contact are capable of causing each other to change, necessarily whatever things have the power of mutually causing change will touch what touches *them*. But why is it that for the most part what touches something touches what touches *it*? Because the things around us, i.e. things that are in <process of> coming-to-be practically all are changed in turn when they cause change. He is right to say 'practically', because neither the image[54] nor the beloved nor, generally speaking, what is desired, while causing <other things to> change, is in turn changed itself.

And things which in this way, while causing <other things to> change, are in turn changed themselves are touched by the things they touch, but not everything that touches is touched by the thing it touches, because not everything that causes <other things to> change is in turn while doing so itself changed. So in those cases where things which cause <other things to> change are not in turn changed by the things they change, they themselves touch the things they change, but the things that are changed do not touch the things that change them. It follows that the celestial bodies touch the things that are here, but these do not touch the celestial bodies, and the slander touches the person slandered, but the person slandered does not touch the slander. The same holds for what is desired and the faculty of desire, where we speak of 'touching' in the more general sense, not in the strict sense. Again it follows that the things that are said in the strict sense to touch touch things which touch *them*, since when they cause change they are in turn changed themselves. But the things which touch, though not in the strict sense, touch things which do not touch them, since they also cause change without in turn being changed themselves.

> **323a27** In these cases it is necessary and indeed apparent that the thing which touches touch the thing which in turn touches *it*. [It is possible, though, as we sometimes say, for the thing that causes change 'just to touch' the thing changed, without the thing which touches touching a thing which touches *it*.]

We should add the word 'as', to make the sentence read 'In these cases it is necessary that the thing which touches touch the thing which in turn touches *it*, as indeed is apparent'. And having said 'It is possible, though, as we sometimes say, for the thing that causes change "just to touch" the thing changed', he adds, 'without the thing which touches touching a thing which touches *it*', as though it were equivalent to 'without the thing which is changed touching the thing which causes it to change'.

> **323a30** But since things of the same kind cause change while being changed <themselves>, it seems necessary for it to be touching something which touches it. [So if something changes something without being changed it will touch the thing changed, but nothing will touch it. After all, we say at times that a man who grieves us 'touches' us, but we do not touch him.]

Things which touch seem to be among the relatives, since that which touches touches something which touches *it*; but he says himself that it is not true in all cases that that which touches touches something which touches *it*. He adds therefore the reason why we think that everything that touches touches some thing which touches *it*. Because, to quote his earlier remark, 'practically all the things we meet', i.e. the things that are about us, I mean those which are in <process of> coming-to-be and perishing, are of this kind – for this reason we think that in fact everything is like this.

He says 'things of the same kind' instead of 'things which have the same matter' (for frequently 'kind' refers to the substratum and matter), and these are what cause change while in turn being changed. Or else 'things of the same kind' simply means 'bodies'. For even if it is not the case that every body while causing change is reciprocally changed by that which is caused to change by it, nevertheless in all cases what is in this way changed itself causes change. So this is why it seems also that everything that touches touches something that touches *it*. However, even if it is indeed caused to change, it is still not reciprocally changed. And for this reason it is not the case either that what touches touches something that touches *it*. So if there were also something which caused changed while being itself unchanged, so much the more would it be the case that it would

touch the thing which is changed without the thing which is changed touching it. And we have said that things which touch in this way, without the things they touch touching *them*, are not touching in the strict sense either. For everything that is relational is asserted 140,1
convertibly. So if something touches in the strict sense of the word, it touches something that touches it.[55]

323a33 This, then, is the way in which we define the notion of *contact* in so far as physical objects are concerned.

For he has said himself that contact is one of the things spoken of in many ways. But contact which is spoken about in the strict sense is <that which occurs> between physical objects. However, he has made mention also of contact in the non-physical sense, that <which occurs> between mathematical objects and <that which occurs> in the case of objects which touch things which do not touch *them*, which do not touch in the physical sense, or rather not even in the strict sense.

<1.7. ACTING AND BEING AFFECTED (1). LIKE AND UNLIKE>

323b1 The next thing to be discussed is acting and being affected. [The views we have received from earlier philosophers are in opposition to one another. Most of them indeed are unanimous in holding this, that like is in every case incapable of being affected by like …]

Having discussed contact, he now discusses acting and being affected. His object is not to distinguish contact from acting and being affected (for he has already made this distinction), but to show how acting and being affected occur. As his custom is, he now sets out first the opinions of his predecessors on acting and being affected, and having refuted these, he introduces his own <views> as follows. The majority of those who have discussed acting and being affected, he says, have held that contrary acts upon contrary, on the grounds that like cannot be affected by like. For if like were affected by like, <a thing> would itself be affected by itself. But if this were so, none of the things which are would be imperishable or eternal. So it is impossible that like should act upon like. For why should it act rather than be acted upon, given that the potentiality <to act> is the same in the agent and the patient?[56] But as it is, what we see is some things having a power of acting and affecting <others> and some a power rather of being affected.

So contrary will be affected by contrary and not like by like. For

even if it seems that like is affected in some way by like – a small amount of fire by a large amount,[57] for instance (for if you bring a lamp into close proximity to a conflagration, it is extinguished) – it is affected not *qua* like but *qua* contrary (for a large amount is contrary to the small), because the matter of the smaller fire is used up by the greater. The majority, as I have already said, maintain that contrary is affected by contrary. Democritus, on the other hand, states idiosyncratically against the others that like is affected by like. For, he says, things that are altogether unlike and different are naturally unable[58] to act on one another or to be affected by one another. For this reason
141,1 sweetness will have no effect on blackness, nor heaviness on sweetness, since these have nothing in common. So it has to be the case that things which act <upon> and are affected <by each other> are not unlike but the same. For colour acts upon colour and flavour on flavour and the rest in the same way.

So like things and not contraries are capable of acting upon and being affected by each other. But I maintain, <Aristotle> says, that things that act <upon> and are affected <by each other> are neither alike nor altogether unlike and contrary. They are not alike, because what is like is in no way affected by what is like. <For if like> were affected in some way <by like>,[59] it would also be affected by itself, since everything is itself like itself. But if so, nothing will be imperishable and eternal. These are reasons why things that act <upon> and are affected <by each other> will not be alike. But neither will they be entirely different and contrary <to each other>. For a line is not affected by whiteness nor sweetness by blackness. So this is the way in which he overturns the opinions of his predecessors, and he then moves on to his own theory.

These, then, are the things that were said by the older <philosophers> about acting and being affected. However, he says, although these people say contrary things, they can probably be drawn together. The reason for this is that each side paid attention only to a part when what was needed was to consider the whole. For in reality things that act <upon> and are affected <by each other> are neither entirely different nor entirely the same, but in one way like and in another unlike and contrary. For it is not any old thing which acts upon any old thing (for heat does not act upon sweetness or upon a line), but contrary upon contrary (for white acts upon black and sweet upon bitter), and contraries have in common both the genus and the substratum. It is obvious, therefore, that things that act <upon> and are affected <by each other> are in one way the same – in respect of the genus and the substratum – and in another way contrary – in respect of species. So those of the older <philosophers> who paid attention to the genus of the things that act <upon> and are affected <by each other> said that like things acted upon and were affected by each other, while those <who paid attention> to the species <said that> contrary things <did so>.

> **323b4** ... since one is no more capable of acting or being affected than the other, [for all the same things belong equally to like things; and that it is unlike and different things which tend naturally to act on and be affected by each other. Even in the case where a smaller fire is destroyed by a larger one it is thus affected, they say, in virtue of the contrariety, since much is the contrary of little. Democritus, on the other hand, puts forward his own view against the others, an odd man out. According to him agent and patient are the same and like, since it is impossible for things which are different and dissimilar to be affected by each other: if things that are different do affect one another, it is not *qua* different but *qua* possessing something in common that this happens to them.
>
> These, then, are the views that have been expressed on the subject. It seems as though those who speak in this way are manifestly saying opposite things; but the explanation of these opposing ways of talking is that each side in effect states only a part when it is necessary to consider the whole. That which is like, in the sense of being altogether and in every respect indiscernible, might well be argued not to be in any way affected by its like. Why, the argument might go, should one be active rather than the other?]

That which acts has to have the power of affecting, that which is affected the potentiality of being affected.[60] But if like things have an equal and indeed identical potentiality, why should either of them be active rather than affected?[61] So given that they have a similar and indeed the same potentiality both of acting and of being affected, they will neither act upon nor be affected by one another.

Should someone say that there is nothing absurd in things that act 142,1
also being reciprocally affected (since it is things which have the same matter which do this), I would reply that being reciprocally affected is not absurd, but that <one should> no more <be affected than the other> is absurd. For even if that which acts *is* reciprocally affected, nevertheless it acts more than it is affected, and even if that which is affected also acts, it is affected more than it acts. But in the case of like things, potentiality for acting or being affected is no more in one than in the other. So like will not be affected by like. And if it is the aim of the thing which acts to change that which it affects into its own nature,[62] whereas like could not be turned into what is like (since it is already even before the conversion that into which it changes), manifestly like will not be affected by like.

323b22 [Again, if something is capable of being affected by what is like it, it will also be capable of being affected by itself; and if this were the case] there would be nothing imperishable or unchangeable [given that like *qua* like is capable of acting, for everything would move itself. The same goes for that which is completely different and in no respect the same. Whiteness could not be affected in any way by a line or a line by whiteness, unless perhaps *per accidens*: e.g. the line might *per accidens* be white or black.]

And in fact to cause change is to act in some way. So if something is affected in some way by itself, everything will in every way be changed in some way by itself and everything will be destroyed by itself.[63] So there will be nothing imperishable or unchangeable, which is impossible. For it was proved in the *Physics*[64] that everything that is changed is changed by something, and that nothing changes itself, unless it is a question of one part <being changed by> another, as in the case of an animal (for change is caused by one thing, the soul, and what is changed is another, the body). And <it was proved> that not everything is changed, but there are the unchanged causes <of change> prior to the things that are changed.

But it does not seem fair to conclude from this that if like were capable of being affected by like and for this reason acted in some way upon itself, it would straight away be the case that nothing would be unchangeable or imperishable. For what if someone should say that there will be something capable both of causing itself to change and of being affected by itself, namely, that whose nature is to be changed and to be affected, whereas that whose nature entirely prevents this will neither be affected by itself, given that it is not in its nature to be affected, nor cause itself to change, being unchangeable? However, he adds for greater precision, 'given that like *qua* like is capable of acting'; for if it is *qua* like that it is capable of acting, it follows both that everything will be affected by itself and that there will be nothing unchangeable nor imperishable.

323b28 Nothing dislodges another from its nature unless both either are contraries or are from contraries.

He finds evidence for the view that nothing acts upon a thing entirely different from itself, e.g. a line on whiteness, unless *per accidens*, in the fact that they do not dislodge each from its nature.[65] For that
143,1 which acts upon something aims at dislodging it from its own nature. But when black is said to act on the line in the white, it is evident that it is *per accidens* that it acted on the line. For indeed even after

the white has become black the line remains a line and is not dislodged from its nature. So it is obvious that black acts *per accidens* upon the line, for the reason that the line happens[66] to be white. So it is in virtue of the fact that the black has an effect on the white existing in the line that it is said to produce an effect *per accidens* in the line. He mentioned the line's being black by way of an example, but you must read 'line' as 'surface'. By saying 'unless both either are contraries' he indicates the extremes, white and black, and by saying 'or are from contraries' <he indicates> the intermediates, e.g. grey.

> **323b29** But, since acting and being affected belong naturally, not to any old thing, but only to things which have contrariety, [agent and patient are necessarily alike and the same in genus but unlike and contrary in species.]

Up to this point he has been setting out the theories of his predecessors, but from now on he sets out his own view. Intending to show that the agent and the patient are in one way like and in another way unlike, he makes use of a syllogism of this sort, taking it as agreed from its self-evidence that agents and patients either have contrariety or are contraries. 'Having contrariety' means the intermediates, e.g. grey when it acts upon white or some other colour; 'contraries' are, for example, white and black. So the syllogism is like this: Agents and patients are contraries; contraries are in some way the same and in another way different (for they are the same in genus but different in species); therefore, agents and patients are in one way the same and in another way different.

> **323b33** For it is natural for body <to be affected> by body, flavour by flavour, [colour by colour, and generally things that are of a given genus by other such things.]

Examples of agent and patients being like. He does not mean that body *simpliciter*[67] is affected by body, but that which has been acted upon. And again, when he says 'colour is affected by colour', what he 144,1
means, clearly, is that which exists in a substratum.

> **324a1** The reason for this is that contraries are in the same genus, [and it is contraries which act on and are affected by each other. So it is necessary that in one sense agent and patient should be the same and in another sense different and unlike one another.]

The reason, he says, why like is affected by like is that agents and patients are contraries, and contraries <fall> under the same genus: agents and patients, therefore, are similar in genus.

> **324a5** Since patient and agent are the same in genus [and like but unlike in species, and it is contraries that are like this, clearly the things which are capable of acting on and being affected by one another are the contraries and the intermediates.]

Having proved that agents and patients are similar in genus by taking it as agreed that agents and patients are contraries, he now intends to prove what he took as agreed, I mean <the premiss> that agents and patients are contraries. And he proves this in turn by taking it for granted that agents and patients are in one way like and in another way unlike; but contraries are of this sort; therefore, agents and patients <are> contraries. But the proof is circular.

> **324a8** And, since it is between these that perishing and coming-to-be (in the wide sense) occur …

Coming-to-be and perishing cover more than acting and being affected. For acting and being affected are instances of coming-to-be and perishing. Nor do perishing and coming-to-be themselves exist without acting and being affected. So having asserted that acting and being affected occur between contraries, he constructs an argument for this *a fortiori*. For coming-to-be and perishing, he says, occur between these: for coming-to-be is from a contrary to a contrary and perishing likewise, as is proved in the *Physics*.[68] So if coming-to-be and perishing <occur> between contraries, there is no coming-to-be or perishing where there is no contrariety. This is so for the fifth element.

145,1 > **324a9** … it is now therefore understandable that fire heats and cold things cool …

For these are contraries, which is why they are also capable of acting and being affected.

> **324a10** … and generally what is active makes the patient like itself. [For agent and patient are contraries, and coming-to-be is to the contrary. So it is necessary that the patient change into the agent, because this is the way in which coming-to-be will be to the contrary.]

He gives an argument for the thesis that the agent makes the patient like itself, having already demonstrated that coming-to-be is change to the contrary and that the agent and the patient are contraries. So the syllogism is constructed like this: coming-to-be, or rather that which comes to be and, generally, that which is changing changes to something which is contrary to itself; but the agent is contrary to that which is coming-to-be and changing. Therefore, that which is coming-to-be and, generally, what is being affected and is changing changes into the agent. The agent, therefore, makes the patient like itself.

> **324a14** We can see a reason for saying that, although they did not say the same things. [For sometimes it is the substratum which we speak of as being affected, as when we say that the man gets well or gets hot, or cold, or any other things of this kind; sometimes, though, we say that what is cold gets hot, what is ill gets well. And both are true – the same thing happens in the case of the agent: sometimes it is the man that we say heats things, sometimes we say that what is hot heats – for there is a sense in which it is the matter that is affected and another sense in which it is the contrary. The one side paid attention to the former and thought that agent and patient must have something in common; the others paid attention to the latter and thought the opposite.]

It is reasonable, he says, to observe that, although those who say that agents and patients are like and those who say that they are unlike do not say the same things, nevertheless both of them have some hold on nature. And he goes on to explain how each of them has some hold on nature. For, he says, even the common usage of people is wont to speak, now in this way, now in that. For we say, he reminds us, sometimes that the person got well, referring to that which got well from the substratum, sometimes that the sick person got well, referring to him on the basis of the affection. And <we say> sometimes that the stone, sometimes that the cold thing became warm; and similarly in the case of agents, sometimes we say that the human being warmed it, when he clearly warmed it up by touching it, and sometimes that the heat warmed it. So since both agents and patients are sometimes referred to on the basis of the substratum and sometimes on the basis of the affection, those who paid attention to the substratum said that like is affected by like (for agents and patients are like in kind, as was said), while those who <paid attention> to the affection <said that> contrary <is affected> by contrary.

But it is clear that when I say that agents and patients have a common matter I am speaking, not of the proximate and ultimate matter (for it is evident that the body which underlies the colour white is mixed in a different way from that which underlies the colour black, and similarly in other cases), but of prime matter, which remains the

146,1 same while being receptive of both hot and cold and of white and black. But if this is so, and it is on account of prime matter that we say that agents and patients have a common matter, why is it not the case that everything is affected by everything and everything acts upon everything – I mean whiteness on sweetness and on sweet-smelling and foul-smelling and on the others in the same way? For prime matter is one <and the same> and common to them all. So I say that it is not enough for things just to have a common matter if they are to act upon and be affected by each other (for on these terms any old thing would be acting upon any old thing, since they all have a common matter[69]), but in addition to having a common matter the forms of agents and patients must be contraries.

This may prompt the following objection. If agents and patients are contraries which also have prime matter in common, why does not the white that is in a horse act upon the black that is in ebony, and in general everything that is black on everything that is white? I say therefore that these are not things which have a common matter, in the strict sense. For neither should the proximate matter of the changing things be taken <as relevant>, nor the prime matter; for in the case of the simple elements it is three-dimensional extension which is their common matter,[70] but in the case of composite <bodies> one should not take the common matter nor the ultimate matter, but the intermediate. I mean something like this: there is a range of mixture within which human flesh can come to be; within this range the flesh sometimes comes out white, sometimes black, when some qualities are intensified or relaxed.[71] This range of the mixture of flesh I call the common matter of the opposites which change into each other in human flesh. And the same applies in all cases. This is why every contrary does not change into every other, since the matter is not in fact common in the way described for the white that is in milk, say, and the black that is in ebony.[72]

So, setting out the whole theory in detail from the start, I say that agents and patients having proximate matter in common are in all cases contraries, but not all contraries are agents and patients. For neither does hard act upon soft, nor above on below, nor in general white on black, nor every contrary on every contrary, but those contraries act upon and are affected by each other which have the power to affect other things (hot and cold above all are of this kind), whilst those things which act and are affected and which have the same proximate matter are in all cases contraries. I add 'proximate' because cold works upon softness. For the soft is made hard by cold, e.g. wax, or by heat, e.g. clay, and obviously it is neither the case that cold is contrary to the soft that is in wax, nor heat to that which is in clay, nor in general that heat or cold is contrary to softness. They do not act on the soft immediately,[73] because they act immediately on

the contraries. For the cold works upon the heat that is in the wax, draws it out and dissipates it, and evidently some wet substance, in 147,1
which the heat actually has its being, is drawn out and dissipated with it. In this way it comes about that the wax is impacted and becomes hard. So the cold works immediately on the heat and because of this on the wetness, and thus finally *per accidens* on the softness.

It is in the same way that, in the case of water, the cold works on it and makes it ice: immediately it dissipates the intrinsic heat that is in it, or even on occasion extinguishes it completely, and draws out and dissipates together with it some substantial wetness of the water; or else, more correctly, by the intensity of the cold crushes together the parts of the water and solidifies them. In this way it makes the water hard and resistant instead of yielding and soft. Similarly also in the case of the clay, the heat quenches the cold that is in the clay, or else changing it to heat and drawing it out and dissipating it, draws out and dissipates together with the heat the wet substance in which the cold actually had its being; and thus what is left becomes hard instead of soft, the heat primarily working on the cold that is in the clay. This is described in detail in Book 4 of the *Meteorology*.[74]

But perhaps what needs to be said is this: the hot and dry substance of the fire works on the cold and wet <substance> of the clay and changes this or draws it out and dissipates it, and in this way makes the clay hard; so the contrary acts immediately upon the contrary. I have added 'immediately' because of this, and 'having the same matter' because of the celestial bodies. For even if sublunary things have their prime matter in common, namely three-dimensional extension, they do not have their more immediate <matter in common> so as to be able to get rid of one of the contraries and take on the other. So celestial bodies too act immediately on the sphere of fire (for it is what they change primarily), and the sphere of fire is affected immediately by the celestial bodies, but nevertheless these are not contraries. The reason for this, as I have said, is that they do not have the same matter. So it was correct to say that things which act upon and are affected by <each other> immediately, provided they have the same matter, are in all cases contraries.

This objection too might understandably be raised to the theory about contraries, namely, how is it that we say that contraries change into each other. For if it is not the case that all contraries act upon and are affected by each other, it is reasonable to conclude that not all contraries change into each other. For the white that is in white lead would not change into the black that is in ebony or in a raven. Not every contrary, then, changes into every contrary. So if the statement which says that contraries change into each other were true, but the white that is in white lead did not change into the black that is in ebony, it would not be its contrary, which is absurd. For in

148,1 universal terms white is contrary to black. So if the universal white is contrary to the universal black, then the particular <will be contrary to> the particular. It seems then that on one side or the other the argument is compelled to reject one of the commonly agreed <propositions, and to concede> either that white is not *simpliciter* contrary to black, given that not every <example of the one> changes into every <example of the other>, or that contraries do not all change into each other.

We say therefore in reply to this that the proposition that all white is contrary to all black is true (for the one produces contrary affections to the other: everything black contracts sight, everything white dilates[75]). And again the proposition that contraries change into each other is true, for they change into those things that are contrary to them in kind, but nothing changes into a contrary which is numerically the same. That is to say, it is not the case that the cold which is in this water becomes the hot which is in this water,[76] or that the black that is in this body becomes the white that is in this body. For generally speaking, change does not admit of being numerically the same, not even in one and the same object. For the whiteness <that is there> before sunburn is not numerically the same as that <which is there> afterwards, not the health <that is there> before illness as that <which is there> afterwards. This has been proved elsewhere.[77]

If, therefore, one and the same whiteness cannot exist even in the same individual, that <which exists> before the change to black and that <which exists after the change> from black, how much more impossible would it be for the white which is in the white lead to change to the numerically identical black which is in ebony? But the white which is contrary in kind <can change> into the black, not into *this* black *simpliciter*, but into black *simpliciter*. So those contraries which have the power to affect things, for instance, hot and cold, when they are affected by each other and change, change immediately into the same form. For example, when water is affected by fire, it is affected, no doubt, by way of coming-to-be and perishing; it changes immediately into the same heat as was possessed by that which acted on it, e.g. that possessed by fire. But when the water which is being warmed has not yet perished, but is, as it were, being altered and is on its way to perishing, then it changes to the same form of heat, not *simpliciter* the one which is immediate and like the agent, but the form which is regarded as being in the range of heat. And perhaps, even when the water perishes, it does not change immediately to a form specifically identical with that of the agent. For first it becomes like air and changes into a different form and not into the heat that belongs to fire; but the air,[78] or the cold that is in the air, changes immediately into the heat that is in the fire.

149,1 And the contraries which lack any power of affecting things, such as white or black, if they were inseparable, like the white of white lead or the black of pitch, would change into those things that are contraries according to species that <are involved> in coming-to-be and perishing (for the white that is in white lead would come to be black because the white lead perished – it would come to be some bit of ebony or pitch or something of the sort). But if they were separable they would again change into things which are contraries according to species within a range. For the white body would come to be black, not however this <particular> black, e.g. that which is in ebony, since it is not even affected by it, but black *simpliciter*. This then is how the proposition that contraries change into one another is true. Besides this, the solidity of matter often comes to be an obstacle to the change; for the black that is in the ink of the cuttlefish often overwhelms the white that is in milk, but not the black that is in ebony. So for change of the contraries the matter too has to be suitable for acting and being affected.[79]

However, someone will say, 'If, in virtue of the ability of the black body, when destroyed, to make the white body different, you say that contraries in this way change into one another, why should it be said that contraries change into one another rather than that everything without qualification[80] changes into everything on the grounds that from the destruction of anything anything else can come to be?' To this then I reply that it was said in the *Physics*[81] that coming-to-be is in all cases from contrary to contrary. For even if we say that the things that come to be come to be from the privation, it is always the case that there is some form in the privation contrary to the one which will exist and from this the change takes place, even if this is without a name and unknown; for it would not be change unless there were some contrariety between that from which and that to which <there is change>.

So if absolutely every coming-to-be is from a contrary, it is impossible for everything to come to be immediately[82] from everything. For an oyster-shell would hardly come to be immediately from earth, but it would be necessary first for it to be mixed in a particular way and to obtain a certain contrariety towards the form that is to come to be, so that in this way the thing that is capable of acting can change the contrary into its contrary. So this is how, even if everything changes in virtue of the shared prime matter which is able to receive all forms, nevertheless it will first lay hold of some <formula of> mixture which possesses a certain contrariety to the form which is to come to be, and is in this way suitable for changing into the contrary.

324a24 The same account has to be accepted of acting and being 150,1
affected as of being changed and causing change. [There are two

senses of 'that which causes change': that in which the principle of change exists is held to be a mover, for the principle is first among causes ...]

After saying what agents and patients are, namely that they are things that are in one way like and in another way unlike, he now investigates how these, I mean agents and patients, are related to one another, whether every agent is reciprocally affected or not. He says, then, that in order to discover this we should see how things that cause change and things that are changed are related to each other. For agents and patients must necessarily be related to each other in the same way as these[83] are. For causing change extends to more things that acting does, since everything that acts also causes change, whereas not everything that causes change acts, if, that is, acting is bringing something to a form and affection.

He stated in the *Categories* that where things are said of something as of a subject, the things which are said of that which is predicated will be said also of the subject.[84] So those things which belong to the cause of change and that which is changed will belong also to agents and patients. Things that cause change are divided into what causes change primarily and what causes change ultimately and proximately. Nothing, he says, prevents the primary cause of change being unchanged in <the act of> causing change, and in some cases this is actually necessary. Ultimate and proximate causes of change, on the other hand, necessarily cause change while themselves being changed.

For example, the soul is the primary cause of change in the animal, but it is not necessary that this should be absolutely unchanged (for it is changed *per accidens* in <the act of> causing change, altering its position together with the body and not abandoning it when it is caused to change <its position>). The <mental> image, on the other hand, remains unchanged while it acts as the primary cause of change in the faculty of desire. In the same way also slander remains unchanged while causing change in the spirited <part of the soul>.[85] For these things are not even changed *per accidens* in <the act of> causing change. The innate heat, however, which is the proximate cause of change in the animal, and the muscles, or any other parts which are capable of causing change proximately in the whole animal, cause change by being themselves also changed.

These <characteristics>, therefore, belong to the cause of change, and belong to it, not in so far as it is a genus, as does, for example, being predicated of a number of things which differ specifically from each other, but *simpliciter*, in so far as it is an object and a cause of change. For this reason they will belong also to the agent. So not everything that acts is also reciprocally affected, but that which acts primarily is sometimes affected reciprocally and sometimes not. For example, medical expertise acting upon the person who is unwell, and in general bodies of knowledge acting primarily upon learners, remain themselves unchanged. But fire acts primarily upon water, yet for all that it is reciprocally affected by it.

The things which act upon things proximately, however, are also reciprocally affected in every case, as in the case of the medicine which acts upon the person who is unwell and is affected reciprocally by that person and in that of the air in the baths which acts proximately upon us and is affected reciprocally by us. To sum things up, the agents which are reciprocally affected are those which subsist in the same matter. This is why science is not affected reciprocally, e.g. medical expertise by the person who is unwell, but the medicine is affected reciprocally: because knowledge has its being in the soul, whereas sickness <has its being> in the living body and acts reciprocally upon the medicine, because the primary substratum is the common matter of all bodies.

He seems, however, to have drawn the comparison between causes of change and agents incorrectly. For when he enquires what sort of agents are reciprocally affected by the patients in course of acting, after saying that what holds for causes of change holds also for agents, he calls attention to certain agents which are not reciprocally changed while causing change by the things that are changed, but merely changed themselves too while causing change and requiring to be changed in order to cause change, but not being *reciprocally* changed <in>[86] causing change. For all the things that cause change proximately are of this sort, e.g. the innate heat or the muscles, and similarly sticks.[87] But this was not what he was enquiring about, but rather whether they are changed reciprocally while causing change.

I say, therefore, that even if the model does not work completely, it is nevertheless to this extent adequate, that as in the case of things that cause change there are some which are not unchanged, so also in the case of agents, where the example of the soul does in fact preserve the analogy. For the soul in causing change is reciprocally changed, even if *per accidens*. And in general, change is distributed not as a genus into species, but as a homonymous expression[88] <is distributed over> different things that it signifies (for <change> does not exist in <just> one category but in several, and there is nothing which is a genus of categories[89]). And this is why it should not be demanded in general that what holds good for change holds good for action too, but it suffices if the analogy obtains at least up to a point.

324a28 And again, the last in the series (in the direction of the thing changed and coming-to-be) [is regarded as causing change.

Similarly with 'agent': for we say of both the physician and the wine that they make people well. Now, whilst nothing prevents the first cause of change being unchanged during the change, and indeed this is necessarily so in some cases, the last in the series is always a cause of change that is itself changed. In the case of action the first agent is unaffected, but the last is also affected. Things that do not share the same matter act without being acted upon; e.g. the medical skill, which produces health without itself being in any way affected by the thing which is made healthy, whereas the food <that is administered>, in acting, is also affected in some way or other – it is either heated or cooled or is affected in some other way in the course of acting. The medical skill is like a principle and the food is like the last <cause of change> which is in contact <with the thing changed>. Of things which are active those whose form does not exist in matter are incapable of being affected; those whose form is in matter are capable of being affected.]

i.e. the last of the things which cause change or which act, not going upwards, but in the direction of the thing being changed or coming-to-be, i.e. the nearest to the thing being changed or coming-to-be, as the food or the medicine is to the person who is unwell. For this is the last in the direction of the patient.

324b6 For we say that each of a pair of opposites has more or less the same matter, which is like the genus …

He adds 'more or less' because the proximate matter of the opposites
152,1 is not precisely the same; for the elements which are to take on the colour white have to be mixed in one way and those that are to take on black another. And similarly in the other cases. And he says that the matter of the opposites is like a genus because there is one matter for the opposites in the same way as there is one and the same genus; and the opposites differ in respect to their substratum no more than they do in respect to their genus.

324b7 … and that which is capable of being hot, [given the presence and proximity of a heating agent, necessarily gets heated. As has been said, therefore, of things which are active some are incapable, some capable of being affected, and the situation is the same in the case of things which are active as it is in the case of change: there the first cause of change is unchangeable; in the case of things which are active the first agent is incapable of being affected.]

The opposites have the same matter, and this is why that which is capable of being hot (i.e. that which is not yet hot, i.e. what is cold), because it exists in matter whose nature is to take on heat as well, is heated when that which has the power to heat approaches it. So it was correct to say that those agents and patients whose nature is to exist in the same matter are reciprocally affected in the course of acting, because it is the nature of the matter of each, being the same,[90] to receive the form of the opposite; <so>[91] when this approaches it is affected by it, as might be expected.

> **324b14** [The things which is active is a cause in the sense of being that from which change begins.] The final cause is not capable of acting (so health is not capable of acting, except metaphorically). [For when the agent is present the patient comes to be something, but when dispositions are present it is no longer a case of coming-to-be, but of already being.]

He says 'metaphorically' in place of 'incorrectly',[92] or else because when that which is capable of acting is present to the body it is a cause of its being healthy, so in the same way is health.[93] And he constructs a proof that health – and the end and the final cause generally – is not capable of acting because of what he goes on to show. 'For when the agent is present,' he says, 'the patient comes to be something, but when dispositions[94] are present it is no longer a case of coming-to-be, but of already being.' It is clear from this, he says, that the final cause is not capable of acting. For when the agent is present the thing which is changing is still coming-to-be something, and is making its way towards the form having not yet received it perfectly (for so long as the agent is acting, the thing which is coming-to-be has not yet received it). But when ends and dispositions are present it is no longer coming-to-be, but already is; so the final cause is not the agent.

So this is why he nowhere says openly that God is the efficient[95] cause of the universe. For since this is what he says the efficient cause is – that which brings the things that are coming-to-be to being by a certain path and <process of> coming-to-be – and <since> God does not act in time or imperfectly, for this reason he does not say that he is an agent, but that, if at all, he should be called a procurer.[96] But he 153,1
is said also <to be> a final cause on account of the fact that all things are drawn to him and desire him.

> **324b17** And forms and ends are dispositions of a sort. [Matter *qua* matter is passive. So whilst the heat of fire exists in matter, were there such a thing as heat existing separately, this would not be affected in any way. This perhaps is not a thing which

could exist separately; but if there were any such things our remarks would be verified in their case.]

Since he has mentioned 'dispositions', by way of explaining what dispositions he means he adds that they are ends and forms. 'Matter is passive':[97] the matter in question is that which is affected and acted upon and changes in respect of the forms. So this is why he says that those forms which have their being in matter will be affected, because their substratum is neither unchangeable nor impassive. Fire and water and the others are like this. But if there were forms which existed separably from matter, these would also be impassive.

324b22 [This is how we deal with the questions,] what, then, is acting and being affected, [to what do they belong, why and how do they belong to them.]

<His answer is> that acting is bringing to an affection, being affected being brought to an affection, and that to act is to make that which is affected like oneself and being affected is being made like the agent. And <his answer to the question> 'To what do they belong?' <is> that <they belong> to opposites, and <to the question> 'Why?' that <it is> because of their possessing a common matter capable of receiving opposite affections. And <to the question> 'How?' that <it is> by the agent approaching and coming close to the patient.

<1.8. ACTING AND BEING AFFECTED (2). EMPEDOCLES AND THE ATOMISTS>

324b25 We must return to the question of how this can occur.

Having said what acting and being affected are, and what agents and patients are, he intends now to say in what sort of way acting and being affected take place. First, however, he again sets out the ways in which his predecessors said that acting and being affected take place and refutes them, and in this manner sets out the true way.

One school, he says, maintained that acting and being affected took place by means of passages.[98] Empedocles too was a member of this school. <In their view> there were passages in bodies in which the agent entered and thus acted upon the patient. And this is how they said that seeing took place; for while there were passages in all bodies, in the transparent ones they were more frequent and arranged in rows, and through them the visual rays passed and in this way struck against the visible objects. But in bodies that were not transparent, because the passages were not arranged in rows, the visual rays were unable to pass through the whole <body> and strike against the visible objects. Perception occurred in the same way in the other senses, e.g. hearing or smelling, with the quality of fragrance or the sound making its way through the passages.

The school of Democritus and Leucippus, on the other hand, said that acting and being affected took place through the void. For the atoms were separated by the void, in which the agent comes to be and transposes <things> by movement or rearranges the atoms and in this way brings about alteration, or else makes <them> grow by adding itself <to them>. Do the passages differ from the void, since those who introduce passages say that there is no void? <Yes,> for <in their view>, while bodies are full of passages, these passages are filled by a further body, e.g. air or some such thing.

He thinks more highly of the school of Democritus than of Empedocles and the others, because they found the sort of principles which would suffice to provide them with an account of all the things which come to be from nature. For those who introduced passages rendered, if you like, by their means an explanation of acting and being affected, and perhaps of some other things too, e.g. mixing (for they said that it was by means of the passages that bodies were mixed; for mixing takes place, he says, when the passages are so related <to each other> as to be the right size <for the one body> to receive the body which is to be mixed in <with it>). However, they were not similarly capable of rendering an explanation of change, in what sort of way change took place by means of the passages, nor of growth; for the body which causes growth will come in and fill the passages, but it will not increase the size. Democritus, on the other hand, and his followers thought up, Aristotle says, the sort of principles which would equip him for providing the explanation of all the things that happen naturally, and these are the atoms and the void.

He goes on to state the hypotheses of Democritus and of the hypotheses of the school of Parmenides * * *. <For the school of Parmenides>,[99] he says, accepted unclear and unproven principles and tried by their means to refute evident truths about which there is general agreement, but the school of Democritus took principles which were taken as agreed on the basis of perception and used them to construct the things that are non-evident. For Parmenides, wishing to prove that there is no plurality in things, nor yet change, which is to fight against perception and to get rid of what was obvious, takes as an axiom <the proposition> that there is no void, which is not something taken as agreed on the basis of perception, but requires a great deal of proof.

Accepting this <as a premiss> he goes on in this way through an argument:[100] If there is no void, there is no movement; but in fact there is no void; therefore there is no movement. So he accepted the hypothetical proposition as agreed, namely that if there were no void

there would be no movement (for movement could only come about through a void), and similarly the minor premiss 'but in fact there is
155,1 no void'. So taking these as agreed, when they are in fact unclear and doubtful, he uses them to disprove something which *is* agreed, namely <the existence of> movement. However, the argument ought rather to be expressed like this: If there is movement, there is a void; but there is no void; therefore there is no movement.

He refutes plurality in the same way. For if, he says, there is division between things, there is a void (for it is only by means of a void that division takes place); but there is no void; therefore there is no division; therefore the universe is continuous and one. Once again, then, he accepted the hypothetical and the minor premiss as agreed when they are not <in fact> evident, and used them to refute propositions that are <generally> agreed.

So Parmenides uses unobvious premisses in this way to refute things that are clearly true,[101] whereas Democritus accepts principles on the evidence of the senses, namely, that division and plurality exist amongst things, as well as movement, and introduces the void on the basis of these, constructing a proof of the unobvious on the basis of the obvious. For if division, he says, and plurality exist amongst things, so does the void; but division and plurality do exist; so therefore does the void – accepting the hypothetical just as Parmenides does, but not this time the minor premiss. For the former said 'But there is no void', which was not obvious, whereas the latter said 'But division and plurality do exist'.

And the same applies to movement: If there is movement, there is a void; but there is movement; therefore there is a void. Here too he accepts the hypothetical in the same way as Parmenides, but not any longer the minor premiss. Parmenides used the same method to introduce the infinite. For if the universe were finite, it would clearly, he says, be the void that was limiting it; but there is no void; therefore the universe is not finite. For these reasons he approves of Democritus on the grounds that he accepted principles which were in agreement with perception, and from them constructed a theory containing hypotheses and fundamental constituents[102] of natural objects which enabled him to give an account of everything which comes to be from nature.

> **324b26** [Some think that whatever is affected is so] by the agent's entering it – the last agent, that is, or the agent in the strictest sense. [In this way, according to them, we see and heat and perceive all the other perceptions.]

For <on the Empedoclean view> the agent enters via the passages and thus acts – not the primary agent, e.g. medical skill, but the proximate <agent>, e.g. medicine or food.

324b31 [Furthermore, things are seen through air and water and the transparent media because they have passages which, though invisible on account of their small size, are frequent and arranged in rows,] and the more so the more transparent <the medium>.

This, he says, is the reason why the objects of sight are seen better through more transparent <media>: because the more transparent <media> have their passages more frequent and arranged in rows.

324b32 Some then dealt in this way with certain problems, [including indeed Empedocles – not acting and being affected only, but also mixing, which they say occurs in the case of things whose passages have the same size as each other.]

That is to say, they accepted principles which were insufficient in relation to all the physical <phenomena>, but were suited to just a few of them,[103] e.g. how acting and being affected take place or how mixed things become mixed; but they cannot explain on this basis how change or division takes place, nor can they deal with any of the other problems.

324b35 Better progress, however, is made by Leucippus and Democritus,[104] and they produce a theory which covers all the phenomena …

i.e. they decided upon the sort of principles which would ensure for them good progress through the explanations of all the things that come to be.

325a1 … adopting a principle which is in accordance with nature. [For some of the earlier philosophers held that the existent is of necessity one and unchangeable.]

This is the reason, he says, why they make good progress with the working out of their theory, namely, the fact that they accept a principle which is both true and in agreement with nature, I mean <the recognition> that plurality and movement exist amongst things. For some of the older philosophers, he says, thought that neither plurality nor movement existed, which was to be in conflict with the obvious. So Democritus did not deny things that had the support of self-evidence, but used these as starting points.[105]

325a3 For the void is not what is,[106] [and things would not be able to move if there were not a void that is separated. And furthermore there could not be many things if that which held them apart did not exist.]

For only the plenum[107] is existent. You should ask whether it is not possible to understand 'For the void is not what is' here as equivalent to 'The void is not one of the things which are', so that the argument is as follows: If without the void it is not possible for there to be movement, and there is no void, then there is no movement.[108] But in the following discussion of Democritus it is no longer possible to take this to be the meaning of 'the void is not what is', as being altogether non-existent, but only in contradistinction to the plenum. For De-
157,1 mocritus posits the existence of the void as not having any substance. After saying 'if there were not a void' he adds 'which is separated'. For it was said in the *Physics*[109] that the void and place and the plenum are the same in their substratum, but different in their relations. For the void is nothing but place deprived of body and the plenum is place full of body, but place is void capable of receiving body or extension containing body.

325a6 Nor is this any different if one maintains that the whole is not continuous [but that, though divided, its parts are in contact, rather than asserting the existence of many things (and not one) and a void. For, they think, if the whole were everywhere divided, nothing would be one; and so there would not be many things either, but the whole would be a void. To suppose, on the other hand, that it is divided at one point and not at another would be to introduce a certain artificiality, as it were, into the account: to what extent, and why, is part of the whole like this and a plenum, and the rest divided?]

He carries on as if it were Parmenides speaking. By 'this' he means, not what has been said already, but what is about to be said. For Parmenides, supposing that there was no void, rejected plurality and movement on the grounds that it was not possible to assert the existence either of division or of movement without a void, whereas the correct theory is that even though there is no void nothing prevents there being both division and movement, objects being divided from, but in contact with, each other without being kept apart by void. Parmenides, therefore, denying this, says that this assumption there is no different from the introduction of atoms and a void.

For is it the case, <Aristotle> asks, that that which is is everywhere divisible or not? For if it is everywhere divisible, not only will the objects <in the world> not be many, but they will not be one either (for if it is everywhere divided nothing will be left, only void). But to suppose that it is not everywhere divisible would be to introduce a certain artificiality; for why is it divisible in one place but not in another? On this account the atomists' theory too will come romping in, which even concludes that void exists.[110]

> **325a12** Again, on this view it follows just as necessarily that there is no such thing as movement.

By the same argument, he says, he proves that there is no such thing as movement, since he will neither admit a void nor allow that there is movement without a void.

> **325a13** These arguments then lead them to go beyond perception [and pay no attention to it, on the grounds that one is obliged to follow where the argument leads. They maintain that the whole is one and immovable, and some of them say that it is infinite; for the finite would end at the void. This school therefore arrives in this way and for these reasons at its conclusions concerning the truth.]

He blames the followers of Parmenides because they thought they need[111] pay no attention whatever to what is evident from the facts 158,1
but only to what follows from the arguments, in which frame of mind they maintained that the universe was one and immovable and infinite.

> **325a17** Moreover, as a matter of argument these conclusions do appear to follow, [but as far as the facts are concerned to hold opinions of this sort seems to be next door to madness; for no lunatic is so far out of his wits as to think that fire and ice are one. At most it is possible to confuse what appears beautiful because one is accustomed to it with what actually is beautiful – there are people who through madness fail to see the difference between these two things.]

It may, he says, be thought that there are these consequences, I mean the universe's being one and infinite and immovable, in so far as they follow from the arguments and the reasoning is persuasive. Nevertheless, this sort of thing, as far as the facts of the matter are concerned, is no different from madness, or rather it is worse than madness. For not one of those who are mad has delusions in opposition to perception, nor does he think fire is the same as ice, but madmen are deluded are about what is fine and not fine and useful and not useful; their madness causes them to select what is useless in preference to what is useful. But the school of Leucippus and

Democritus said things that were in agreement with perception and did not deny either plurality or movement.

> **325a25** [Leucippus, however, though he had theories which said things which were in agreement with perception and would not deny coming-to-be or perishing or change or the plurality of existing things.] Having conceded these points to the phenomena, [and, to those who argue for the One, the points that there would be no movement without a void and that the void is what is not, he says also that nothing of the existent is non-existent: for the existent in the strict sense is a plenum. Nevertheless, what is like this is not one, but infinite in number and invisible on account of the smallness of its bulk. These bodies move about in a void (for a void exists), and by coming together they produce coming-to-be, and by being dispersed perishing.]

So Leucippus' views are to this extent in harmony with the phenomena. <Aristotle> then accepts <certain doctrines> from those who reduce <everything> to one and concurs with them in holding that there is no movement without a void and that the void is what is not, i.e. not a plenum, nor any substance or solid object, and in this way argues for the consequences of this, I mean <the theory> that there are atoms and a void. For if the things which are are many and divided, and it is impossible that division should occur without a void, then there *is* a void. So in this way he gives an argument for there being a void, and in turn in the following way for there being atoms: Each of the things which are *is* in the strict sense; and in that which is there is nothing which is not; therefore there is no void in the things which are (for the void is something which is not). But if there is no void in them, and it is impossible that division should occur without a void, then it is impossible that they should be divided.

> **325a32** But they act and are affected where they happen to come in contact with each other ...

i.e. by means of the void, for it is with this that they are in contact
159,1 with each other. For when Democritus says that the atoms are in contact with one another he is not speaking of contact strictly understood, which is what occurs when the surfaces of the things which touch fit over each other, but what he calls 'contact' is atoms' being close to each other and at no great distance from each other; for in all cases they are kept apart by the void.

325a34 [… (for they are not one here)], and being placed together and interlocking they generate.[112]

Being placed together with one another and interlocking the atoms, he says, generate the affections, as has often been said. By 'placing together' and 'interlocking' he means position and arrangement.

325a34 But from what is truly one a plurality could not come to be, [nor one from what are truly many: this would be impossible.]

Neither, he says, would what is strictly one make a plurality, given that plurality comes about by division and division takes place by means of the void (so that which is strictly one, having no void in it, would not be divided), nor would one come to be from what are in truth many; for neither can the atoms be aggregated, since they are kept apart by void and cannot be made into one.

325b5 [However, just as Empedocles and some of the others hold that things are affected by way of the passages, so Leucippus maintains that all alteration and all being affected comes to be in this way, the disintegration and perishing of things coming-to-be by way of the void – and similarly also growth, solid bodies slipping in through the gaps.] Empedocles is bound to speak in more or less the same way as Leucippus does.

Empedocles' theory, he says, comes round to the same thing as that of Democritus, in so far as Empedocles' premisses necessarily lead him to introduce atomic bodies. For if he posits that it is through passages that acting and being affected take place, it is absolutely necessary that when the agent enters into the passages the bodies that contain the passages should be unaffected and undivided. For if they were affected, since he says that acting and being affected does not take place except through passages, they too would have to have passages, and so too would the bodies containing *those* passages. So either the whole will be passage, or, if this is impossible, there have to be some ultimate unaffected and undivided bodies. And this is just what Democritus said. So Empedocles will be compelled, so far as his own premisses go, to say the same things as Democritus and Leucippus.

325b6 A number of solid bodies exist, and are undivided, unless 160,1
there are continuous passages everywhere. But this is impossible, [because there would be nothing else solid over and above the passages, but everything would be void. So the things which are in contact are necessarily undivided, and what is between

> them is void, and this is what Empedocles calls 'passages'. And this is how Leucippus too speaks of acting and being affected.]

Empedocles, he says, is bound to say that there are some undivided solids, since there are not continuous passages everywhere in a body. For this is impossible, for the whole body would be passage and void. So if this is absurd, the parts of the body that are in contact will have to be undivided solids, and what is between them will have to be void: and this is what Empedocles called 'passages'. The followers of Leucippus too posited the same theses, that though atoms were in contact with one another they were separated by the void, through which acting and being affected took place, except that Leucippus' followers were not using 'contact' in the strict sense, as has been said above.

> **325b12** This, then, is more or less how the ways in which certain things act and others are acted upon were described.

Having set out the ways in which Empedocles and the followers of Democritus supposed that acting and being affected took place, as between these <Aristotle> gives his approval rather to the followers of Democritus, as he has already done, on the grounds that they posit the sort of principles which are sufficient for them <to explain> coming-to-be and perishing as well as alteration and the other natural changes, Empedocles being unable by his own principles to give accounts of all these things. For in the first place he says that the four elements are incapable of being affected and that their specific differences are the affections, and so is unable to preserve alteration. And although he says that there is such a thing as coming-to-be of composite objects as a result of the aggregation of the elements, he can no longer speak of a coming-to-be of the elements themselves, if he does not posit any more simple things as elements of these themselves. But the elements are seen to perish and come to be.

The followers of Democritus, on the other hand, themselves say too that all composite things have their coming-to-be from the four elements and give an account of the coming-to-be of the elements themselves too from their own <brand of element>; for they come to be and perish by way of the aggregation and segregation of the atoms. And they retained alteration, <interpreting it> by means of the transposition and reordering of the atoms; for since they did not find a home for the affections in the atoms, but said that <the atoms> are incapable of being affected, they maintained that it is because of their position and arrangement that the affections seem to come to be.
161,1 Empedocles, however, having located the affections in the elements,

since he maintained that they are unchangeable, is unable to retain alteration <in his system>.

'This is more or less how ... were described', i.e. by the earlier philosophers. And 'the ways' in question were two: that which appeals to the passages and that which appeals to the void.

325b13 [We are clear, then,] about these philosophers ...

i.e. the followers of Leucippus and Democritus, as is clear from what follows.

325b14 [... and what they say,] and it clearly follows in reasonable agreement with the postulates they use.

Their further remarks, he says, which contain the account of acting and being affected, follow from, and are the consequence of, their principles.

325b15 Less so with the others, [for example Empedocles: it is not at all clear on his theory how there will be coming-to-be and alteration.]

In the case of the followers of Empedocles, he says, their further remarks – i.e. those concerning coming-to-be and perishing and alteration and acting and being affected – follow less well[113] from their principles.

325b17 For the former the primary bodies are indivisible, [differing only in shape, the primary things out of which <other things> are composed and the ultimate things into which they are decomposed. In Empedocles' case it is clear as regards other things, until you reach the elements, that they have coming-to-be and perishing; but as regards these elements themselves it is not clear how the mass of them as it is piled up comes to be and perishes.]

For the followers of Democritus, he says, the primary bodies, from which as primary everything comes to be and into which as the ultimate <bodies everything> is dissolved – these are the indivisibles, because it is into these that the four elements are analysed. But for Empedocles the analysis of all other bodies, i.e. the composite ones, goes as far as the four elements, because for him the four are primary elements. This is why for him the aggregation of the elements would provide an account of the coming-to-be of the composite <bodies>, but he has no means of saying in what way the elements themselves come to be and perish, since he does not speak of any simpler elements of the four elements themselves. The 'piled up mass' is this: that which is piled up from their coming-to-be, e.g. water or air or fire, which as 162,1 they come to be little by little pile up and grow, as the water <that comes> from rain and the air <that comes> from evaporated water or the fire <that comes> from fuel that has been ignited.

> **325b24** [Nor is it possible for him to explain this unless he is prepared to say of fire,] and equally of all the others ...

i.e. water and air and earth.

> **325b24** [... that they have their own elements,] as Plato writes in the *Timaeus*. [This account <of Plato's> is to this extent different from that given by Leucippus, that in the latter solids, while in the former planes are regarded as the indivisibles ...]

For Plato says that the surfaces are elements of the four elements too, and these are equilateral and right-angled triangles.[114] The pyramid he does not call an element of fire, but a seed,[115] and similarly for earth the cube, and in the same way for the others, for water the octahedron and for air the eicosahedron; for[116] he earmarked the dodecahedron for the whole cosmos. But the triangles, i.e. the primary elements, he said were indivisible, since these were no longer divided into other, more primitive, things. So the views of Plato differ in this respect from those of Leucippus and Democritus, he says: that the followers of Leucippus said that the indivisibles, which he also made the primary elements, were solids, but Plato said that the indivisibles were surfaces; and Democritus said that the atoms were shaped by an infinite variety of shapes, Plato by a finite number, or rather by one. For the primary elements are triangles, and the things which are the first to be formed from the elements are defined by a finite number of shapes (five, in fact): the pyramid, the cube, the octahedron, the dodecahedron, the eicosahedron.

> **325b27** ... and in the latter <account> each of the indivisible solids is bounded by an infinite number of shapes, [in the former by a limited number ...]

What he means is not this: 'Each atom has an infinite number of shapes' (for each has one), but 'The atoms have an infinite number of shapes divided between them, one for each atom'. Or else it should be understood in the opposite sense: 'And in the latter <account> each of the shapes is allocated to an infinite number of indivisible <bodies>,' i.e. each shape occurs in an infinite number of atoms, e.g. the spherical, the cubic, and each of the others, so that both the shapes and the atoms allocated to each shape are infinite in number.

325b28 ... whereas both say that they are indivisible and defined by their shape. 163,1

By 'defined' here is meant 'each of the solids is circumscribed by one shape'.

325b30 [Out of these we get the comings-to-be:[117]] for Leucippus there would be two ways, by means of the void and by means of contact (for it is here that each is divided).[118] [But for Plato it only happens by contact, for he says that a void does not exist.]

These remarks seem to be incompatible with what he has said earlier. For in those <remarks> he says that being affected takes place in this way, as a result of dissolution through the void, and similarly growth and coming-to-be and perishing and the other changes. Now, however, he says that there are two ways of coming-to-be, one by means of the void and one by means of contact. So it looks as though here he is using 'coming-to-be' in the wider sense, both for the true coming-to-be, I mean that of substances, and for that of accidents, and the one 'way', by means of the void, belongs to coming-to-be, and the other, by means of contact, to alteration. For when the atoms come in contact with one another and strike against each other their position and arrangement are changed, and in this way they produce alteration; and they obviously come into contact with one another through the void. So the former are not opposite to the latter.

But 'two ways' might indicate, not that coming-to-be takes place in one way and in another way, now in this, now in that, but that there are two ways through which each coming-to-be <occurs>, and that each of the things which comes to be requires the two ways for it to come to be, namely, that of the void and that of contact, because contact itself takes place by means of the void, whether you mean contact in the strict sense or that which is like coming close.

No doubt in the earlier remarks as well he meant that acting and being affected both take place by means of contact. What he said was this: 'And by coming together they produce coming-to-be and by being dispersed perishing; but they act and are affected where they happen to come in contact with each other.' Clearly it is not only the things which are altered that act and are affected, but also the things which come-to-be.

In this way, then, there were for Democritus, he says, two ways of coming-to-be, but for Plato <just> one, that by means of contact. For

Plato says there is no void, but the surfaces, touching each other in
164,1 the strict sense of 'touch', produce the pyramids or the cubes, and the other shapes that are solid, and these in turn by touching each other produce the elements.

325b33 We have spoken about the indivisible planes in a previous work.

For he refutes them, both in the *On the Heavens*[119] and in this work,[120] on the grounds that it is unreasonable to pursue the analysis as far as plane surfaces (for by parity of reasoning the analysis should continue right up to points), as well as the other things that are said.

325b34 We must leave for the moment the more thorough consideration of the consequences of the indivisible solids.

Aristotle has devoted a whole treatise, the *On Indivisible Magnitudes*, to the refutation of the theory of atomic bodies, as well as setting it out in the *On the Heavens*; and we should say how many arguments and what sort of arguments he uses in the *On the Heavens* to produce his refutation of the theory. So he sets out in the third book some six arguments, and one in the fourth adapted to the theory of heavy and light.

The first of the arguments, then, is that the authors of the atomic theory do not posit fundamental constituents suitable for someone who is to know <the truth of the theory>, since they say they are infinite in number. For the infinite is by its own nature unknowable, but science is of finite things and knowledge is finite. But this argument is not directed against the indivisibility of the atoms, but against their infinite number.

The second argument is that the proponents of the theory abolish the most exact of the sciences, i.e. the mathematical <sciences>. For the latter[121] divide in two everything possessed of size, including that which is conceived in the mind, but for the former[122] this does not hold even for natural bodies; for that which is composed of an odd number of atoms cannot be divided in two.[123] So they are in conflict with the most exact of the sciences.

Thirdly, they abolish self-awareness and the affections.[124] For, as has been said several times, when some sharp instrument pierces the body, it must of necessity either move in the void[125] or <touch>[126] the atom. <If it moves in the void, it will not be perceived; if, on the other
165,1 hand,> it touches <the atom>, once again it will not be perceived on account of their[127] being incapable of being affected. So perception is abolished and nothing is perceptible. For if 'perceiving' means being affected in some way by the perceptible <objects>, but there is no such

thing as self-awareness or affection, then there could not be anything perceptible either, in the strict sense. But if it were to be said that nothing was perceptible by nature, but <only> in virtue of position and of a certain type of relationship to the senses (for perceptible objects seem to exist in virtue of the atoms' striking against <the sense organs> in one way or another), note that this argument was successfully refuted by Plato in the *Theaetetus*.[128]

The fourth argument is that they abolish continuity and division; for nothing will be continuous, and, because of that, not divisible either. However, it is clear that bodies are continuous from the fact that when one part is affected practically the whole body instantaneously has awareness, since the affection is transmitted through it owing to its continuity. If, however, the body were not continuous how, when a part is affected, e.g. if you stub your toe, is the awareness going to spread instantaneously through the whole without a break? For if ten of us stand next to each other in a line, and then one is wounded, awareness does not come to be for the others, because they are not continuous <with each other>, but divided.

The fifth argument is that they abolish coming-to-be and perishing. For how would air come to be from water when it evaporates? No doubt they will say <that it happens> as a result of the emergence of the atoms from which, when they have been aggregated, the air is composed. But if this were so, it ought never to be the case that the whole of the water was evaporated, but that some of what was being evaporated was always left behind. For the same atoms are not capable of producing all the elements, but different ones produce different <elements>. So if it is not in virtue of a conversion of the substance of the water that air comes to be, but because the atoms of air are extruded from it, it ought never to be the case that the totality of the water evaporates. So either there is no such thing as coming-to-be at all, or else the atomic theory is false; for it is impossible on this hypothesis[129] for the coming-to-be of the elements from each other to be retained. Nevertheless, they themselves maintain, he says, that the elements come to be from one another. So they contradict themselves, saying on the one hand that the elements come to be from one another, and on the other hand, so far as their own hypotheses go, making <their> coming-to-be from one another impossible.

The sixth argument against these <philosophers> is that if the atoms differ from each other in their shapes and the shapes are infinite in number, then the <different> natures of the atoms will be infinite in number. But if there are infinitely many natures, there will also be infinitely many <types of> movement and infinitely many <natural> places; but as it is, neither the <types of> movement nor the <natural> places are infinite in number. And if it were said 'But they said that <the atoms> are all heavy,'[130] as he also said, 'and that

they all move downwards, and that the upward movement comes to be as a result of certain atoms, e.g. the spherical ones, being squeezed
166,1 out', we will reply that there are some things which are moved by some force or other against nature towards a place which is unnatural <for them>, and the larger they are the slower they will move according to this unnatural movement. For if someone throws a large piece of earth upwards it will move more slowly than a small lump. So with fire too a large quantity ought to move upwards more slowly and a small quantity more quickly, which is the opposite of what we actually see; for the greater the amount of fire, the faster is its upward movement.

These are the arguments he sets out in Book 3.[131] In Book 4 he says that those who postulate the atoms and the void will have no explanation to give for the fact that some things are heavier or lighter than others. For why are bodies divided into those which are heavier and those which are lighter? For it must be either on account merely of the plenum or on account merely of the void or on account of both. For example, attributing it solely to the plenum enables them to say that things which have more of the plenum are heavier while things which have less of the plenum are lighter, with the void contributing nothing; but attributing it to the void results in the things which have a larger share of the void being lighter, those with a smaller share heavier, whilst explaining it through the proportion of void to plenum allows us to say that those are lighter in which the ratio of the void to the void is greater than that of the plenum to the plenum, the others being heavier.[132] So that, say, where there are two bodies, the void in the one will have a higher ratio <to the void in the other>[133] than the plenum <in the one>[134] does to the plenum in the other, so that the amount of void <in the one> might be, say, double the amount of void in the other, but the amount of plenum <in the one> one and a half times, or equal <to, that in the other>. So if it were due to the plenum only that some things were heavier, some lighter, a small amount of earth would have to be lighter than a large amount of fire; for the plenum is greater in the large quantity of fire than in the small quantity of earth. If, however, <it were due> to the void only, the large quantity of earth would have to be lighter than the small quantity of fire; for there is more void in the large quantity of earth than in the small quantity of fire. But if it were due to the proportion between the two of them, and those things were lighter in which the ratio of the void to the void was greater than that of the plenum to the plenum, clearly those in which the ratios of the void and the plenum are equal will be equally light and heavy. So a large amount of earth would have to be just as light or heavy as the part of itself; for the ratio of the void to the plenum is the same in the whole as in the part. So the stone which weighs a talent and the small quantity of earth are equal in weight to the most exiguous parts of themselves, which is impossible.

These, then, are the arguments which occur in *On the Heavens*. Here, however, he presents certain other arguments.

> **325b36** But, by way of a short digression, we may point out that each of the indivisibles has to be said to be incapable both of being affected (for it is not possible to be affected except by means of the void) and of producing any affection <in anything else>, [for they can be neither hard nor cold. But this, at least, is absurd, to attribute heat to nothing except the spherical shape; for it would surely be necessary for cold, its contrary, to belong to some other shape. And it would be absurd too to suppose that they possess these qualities, heat and cold I mean, but not heaviness or lightness of hardness or softness. Now Democritus asserts that the indivisibles vary in weight in accordance with their size; so they must clearly vary also in their degree of heat. But if they are like this, they cannot fail to be affected by one another: for example, that which has a slight degree of heat by that which has a degree of heat greatly in excess of it. Again, if hard, they can also be soft; but a thing is called soft precisely because it is affected in some way, for it is that which tends to yield which is soft. 167,1

Those who posit the atoms are bound, he says, to maintain that they are incapable of being affected. For, since they say that being affected does not occur otherwise than by means of the void, and there is no void in the atoms, neither, in that case, do they have any affection. But if the atoms are incapable of being affected, they will not affect themselves; for the affection which they do not possess could never be transmitted either to another or to themselves. And if they do not affect anything, there will be nothing which acts and nothing which is affected either – an absurdity which conflicts with what is manifestly the case. So nothing will be hard after being soft or cold after being hot or anything else of this kind.

Nevertheless, he says, their own arguments conflict with one another. For they say that the spherical atoms are hot, and furthermore that the atoms which surpass the others <in size> are heavier. So, he says, if the larger atoms are heavier, it is obvious that the larger spherical atoms are also hotter. And being hotter and heavier, how can they avoid acting upon and being affected by each other? For that which has a small degree of heat is affected by that which is very hot. So, to the contrary, they *will* act upon and be affected by each other. But it is absurd to say that the spherical atoms are hot, but not also to ascribe the contrary of heat, namely, cold to some other shape. It is absurd also to say that heat and heaviness belong to the atoms, but to deny that their contraries, and the other affections in general, belong to them; for if hardness, he says, belongs to the atoms, why should not softness also belong <to them>? But if softness belongs <to something>, and what is soft tends to yield, and a thing yields in virtue of being affected in some way, then in this respect too they will not be incapable of being affected.

326a14 Again it would be absurd if nothing belonged to them except shape …

We must set out by way of division all the things that could have belonged to the atoms, so that by refuting these we may demonstrate that the atomic hypothesis cannot stand up in any way or by any means. So, whilst the Philosopher himself does not go
168,1 through completely all the branches[135] of the division, let our listing of them, on the other hand, be complete. The atoms, then, must either have no affections or have some. And if they were to have an affection, either they would each have one, or they would each have more than one, or several of them would have one, or several of them would have several; and let the branch <of the division> which posits that several do be taken to include the case where all do.

So if they had no affection, the things which are composed of them would have no affection either. For how could they give what they do not have to the things which are composed of them?[136] But if the things which are composed of them have no affection, perception and perceptible objects will be abolished. For perception is the name for being affected by perceptible objects, and affections are what these are. So there will be nothing whose nature is to be perceived.[137] But if there are no perceptible objects, there will be no perception; for these belong to the category of relation.[138] The fact that perceptible objects have their being in relation to perceptions[139] is proved by Plato in the *Theaetetus*[140] and by Aristotle himself in many places. So the result would be that we would get rid of both the perceptible object and perception. It is impossible, then, that the <atoms> should have no affection.

But if the <atoms> have <affections>, but have <just> one each – and by having just one each I mean no one of them having the same as any other, but each one <an affection> peculiar to itself and not available to the rest – if, therefore, each has <just> one affection, they will not be of the same nature. For if the one is not by nature capable of receiving an affection of the other, they will not be of one nature. And yet they say that they are. But if the <atoms> were not of the same nature, they would not act upon or be affected by one another. So acting and being affected will thus be eliminated, and the affections in the atoms> will be in vain if they neither do anything nor are affected at all by each other. But nothing is done by nature in vain; so it is not possible that each one should have just one affection.

But if each had several – and by each having several I mean, for example, if there were a thousand atoms, each of them having, say, some five affections, not, however, the same as any of the others, but each having its own peculiar heap of affections. Then if each has several in this way, I will demand 'How does it have these, and according to what definition <of "have">'? Either it will be in terms of being composed of parts, or else it will be in the way form is in matter. By 'in terms of being composed of parts' I mean, not that the atom is composed of parts that are actually divided, but that the affections exist in each on account of such and such an attunement of the parts – parts, that is, that are objects of thought. So if the affections existed in them on account of the composition of parts, it would result in being affected in respect of the very same thing in respect of which it acts. For example, if the atom had both heat and wetness, and then acted in respect of the heat and was affected in respect of the wetness, since that in respect of which it is hot is that in respect of which it is wet, in being affected in respect of wetness it would be affected also in respect of heat. So while acting in respect of the heat it would be affected in respect of heat, which is illogical; for while it is heating it will be heated by the heat with which it is heating. So it will not have the affections on account of the composition of its parts.

The same absurdity would seem to follow both for those who say that forms are products of the mixing, e.g. the medical scientists,[141] and for those who say, not that they are products of the mixing, but that the forms supervene from outside on the suitability of the mixing from the creation as a whole,[142] as indeed the correct account has it.[143] For suppose that the forms were products of the mixing: then, since the outcome of the first mixing which measures ingredients in accordance with such a formula – I mean <a mixing> of dry things with wet and hot things with cold – is more than one quality (for colour is like this, and taste and softness and hardness and such things), obviously the existence of sweet and red and wet and hot, etc., will all depend on the same formula. If, therefore, they all depend on the same formula, something which is affected in respect of one of them will be affected in respect of the others too. So if it acts in respect of one but is affected in respect of something else, it will be affected in respect of the very same thing in respect of which it acts; for both depend on the same formula. If, on the other hand, the forms are not products of the mixing but are brought about from outside on the suitability of the mixing, clearly once again different affections supervene on the same suitability. But if different affections supervene on the same formula of suitability, we shall meet with the same absurd consequences, if something acts in virtue of one and is affected in virtue of another. For once again it will be affected in respect of the very same thing in respect of which it acts. The same absurdity, therefore, follows from the correct account as well.[144]

It would, then, be worth investigating this problem in its own right, the problem, namely, of how different forms are produced by the same mixing, or, as the more correct account has it, <how> different forms supervene on the presence of the same suitability; for it is evident that the forms are not simply products of the mixing. For let hot and cold or dry and wet or heavy and light be products of the primary mixing: where <in this case> will the fragrant and the malodorous and the infinite number of subspecies[145] of these come from? And whence the multitude of tastes and colours and the substantial forms themselves, I mean the form of flesh and bone and the infinite number of <other> forms? And if in the case of the homoeomers different forms depend on the same formula involved in the primary mixing, e.g. sweet and yellow and viscous in the case of honey, why should it not be the case that when one is affected the other is also affected with it? For the colour of the honey is affected and from being yellow becomes white, but the sweetness, as like as not, is not affected at all. But obviously if the mixture of the honey had not changed, the colour would not have changed. How, then, is it that the taste did not change as well, given that even at the beginning their existence was determined by the same formula of mixing? Similarly when the wine went sour, the taste changed but the colour did not. And yet if the mixture was affected they ought both to have been affected together, since the existence of both of them was determined by the same formula of mixing. But <even> if in some circumstances the qualities are affected together, as when wine is boiled – and I am speaking of what is called 'mash' or 'sun-scorched'[146] (for in the boiling the colour is changed together with the taste) – nevertheless, in those cases where this does not happen, the theory calls for examination and gives rise to problems which are not easy to solve.

How, then, shall we reply to these points? Each quality is reckoned to be within a certain range, and its being is not point-like.[147] For whiteness exists within a range (for the moment something stops being ultra-white it is not the case that it is now no longer white at all), and similarly sweetness is reckoned to exist within a certain range (for the range of sweet things is extensive), and the same sort of thing obtains for them all. So if the qualities have a certain range, clearly the suitable mixtures in which it is their nature to inhere will have a certain range. However, there is some ultimate formula for a mixture in which it is no longer its nature to inhere, but as soon as the mixture reaches that point, the whole nature of the quality is altered. For instance, to put the argument to work by means of an example, let, say, ten parts of hot and an equivalent number of cold and of dry and wet be <what determines the mixture> in which it is the nature of the form of the most intense sweetness to inhere. If it is reduced by one part of one side of the balance, it will not be sweet to that degree, but it will not be the case that it is not sweet at all – rather that it is less sweet. If, however, it is reduced by five parts (let us suppose that sweetness is retained up to this point), evidently the form will disappear lock, stock and barrel.[148]

So when different qualities follow on the same mixture, they will not all inhere in accordance with exactly the same formula as one another, but sweetness, say, will inhere in its perfect form, whilst whiteness or yellowness or viscosity <will> not <inhere> in accordance with the perfect formula for themselves, but frequently in accordance with a less intense formula. So for this reason if the mixture is modified a little, while the sweetness is not totally modified, the whiteness is often changed completely, since the original mixture was near that particular and point-like[149] mixture in the presence of which this sort of colour no longer naturally occurs. For obviously, even if the modification of the original mixture is slight, no quality remains unaltered from its original condition. For neither does the sweetness of the honey remain the same when it is compacted by boiling or pounding, nor does the colour of the wine when it turns sour, but where perception is concerned, and in so far as it is within a given range, it seems to remain the same. But since, when the honey has been made drier by boiling or pounding, it is natural for the sweetness still to remain, as in a given range of mixture * * *.[150] It is possible to say that in the honey the spirit[151] which has been taken up into it is responsible for the white colour and not that it belongs *simpliciter* to the nature of the honey; but since it is manifestly the case with vinegar and some other things that the colour, in so far as it is within a certain range, remains, whilst the taste <does> not <remain> even roughly <the same>, this persuades us not to stick doggedly to the suggestion even in the case of honey.

So much for the question of why qualities produced by the same formula in the primary mixture are not affected together with each other. But let us return to our original topic. If affections in each of the atoms[152] vary in accordance with the combination of parts, the same thing will both act and be affected in the same respect. But perhaps it will be said that the atom does not derive its affections *simpliciter* from the combination of its parts so that they are all determined by the same formula, but that it comes about that the matter has different <affections> owing to each part having its own peculiar affection. In that case our reply will be, first, the same argument in the case of each of the parts which we gave when we dealt with the hypothesis that each <atom> had <just> one affection; and next, that if different parts have different affections, clearly this is due to the fact that each of the parts is of a different nature, given that it is not the nature of all <the others> to receive the affection of the one part. For if this is its nature, either it will at some time receive it and will change, in which case they[153] will not be changeless, or, if it does not receive it, it will have this potentiality in vain.

And if the parts are of a different nature, what was it that joined them together? And how were things which were in every way separate joined together? And the atoms will no longer be homoeomerous, nor primary and fundamental constituents,[154] but <that will be true rather> of those things out of which these are composed. And if <they exist> as form in matter, their account will coincide with our own, correct, account. So if there is a common underlying substance of the atoms, each will be capable of receiving the affections that are in the other atoms. In that case they will act upon and be affected by each other. For it has been demonstrated that things which have a common matter are capable of acting upon and being affected by one another, having, as they do, contrary affections. So if all the atoms have one side only of the pairs of contraries, nature will limp[155] and they will exist in vain. But if the contraries were to exist in the same atom, it would be absurd for the contraries to exist at the same time. And if the opposing members of the pairs of contrary affections had been shared out, given that they have a common substratum, it would be impossible for them not to act upon and be affected by each other. For the nature of contraries is recognised <precisely> by means of this <feature>.

And if more than one atom were to have one affection, by 'more than one having one affection' I mean all having one or having one by classes,[156] e.g. ten having white and another ten having black. If the first <of these is what is meant>, all of them would have to be of the same nature and there could be nothing to distinguish one from another, whereas as it is, there are an infinite number of types having different qualities; if the second <is what is meant>, if in all the classes one side only of the pairs of contraries belongs, again nature will limp and acting and being affected will be abolished (for it is the contraries which act and are affected).

But if the atoms divide up between them the contrary affections, it is evident that they will act upon and be affected and change to each other because of a common substratum, or else the contraries, neither acting nor being affected, will exist in vain. But if more than one <atom> were to have more than one <affection> – by more than one having more than one I mean either all having all or different classes again having different classes,[157] e.g. ten having so many affections and another ten others – again the same applies in both cases, namely, that if (i) the same <atoms> have both of the contraries, it will be absurd for the contraries to exist at the same time in the same thing, but if (ii) all have just one side of the pairs of contraries, nature will again limp. And there is a special objection to the first suggestion, namely, that all things could not have the same nature, and as to the second, we shall repeat what we said both about the suggestion that there is more than one in each (for the same applies here too) and about the suggestion that there is one in more than one.

Again, it would be absurd both if nothing belonged to them except shape ... And yet shape itself is a quality. What, then, is the rationale for allotting this kind of quality but no other? And having said that the suggestion that the atoms have no affections leads to absurdity, he is silent about the absurdity as something manifest; but we have spelled it out.

> **326a16** ... and if something did, but only one, hardness here, heat there, for example. [For there would not be some one nature which they had.]

He sets out the second branch,[158] which maintains that to each atom belongs <just> one quality which belongs to no other atom; but it is absurd for the <atoms> not to be of the same nature.

> **326a17** But it would be equally impossible for them to have more than one <quality apiece>.

Because it is not possible for each atom to have more than one affection either. And having set out the division as far as the branches we are now dealing with, he dismisses the ones that remain of those which we set out.

> **326a18** For, since each one is indivisible, that in which it has these affections will be one and the same; [so that if it is affected where it is cooled, there it will act or be affected in some other way too.]

If we take him to mean by 'indivisible'[159] something like 'not being by nature divided', his conclusion does not seem to be drawn entirely logically. For it is not the case that if it is not by nature divided, it immediately follows that the affections will exist in one and the same thing (for it will be possible for different ones to be in different parts). But <all will be well if he means> 'indivisible in terms of potentiality';

for it would not be logically possible for different potentialities to be allocated to different parts. For those who postulated the atoms did not say this – and if someone did say it, he no longer adhered to the conception and hypothesis of the atoms. For it would no longer be they which were primary and fundamental constituents,[160] but those <parts would be primary> (I mean these parts, different in substance, out of which even the <atoms> are composed; and things which have a share in others' potentiality are different in substance).

Nor would they strictly speaking still be atoms. For they would be distinguishable by the different potentialities, even if not spatially <distinct>. So since the atoms must also be indivisible in potentiality, it will entail their having several affections according to the same potentiality. So when one is affected the rest must be affected with it.[161] And if it were to have both active and passive affections, since both would belong <to it> in virtue of the same potentiality, it would be being affected in respect of the very same thing in respect of which it would be acting, as we said earlier.[162]

> **326a20** And in the same manner in the case of the other affections too.

Since he tried out the argument in the case of cold, he says that it would be the same in the case of the other qualities too. For that which acts in respect of heat will act too in respect of looseness of texture, if its texture is loose. And if it is being affected in respect of wetness while being hot and is acting in respect of heat, it will at the same time be affected in respect of <heat>, for it is wet and hot in respect of the same thing.

> **326a21** For this consequence will follow in the same way both for those who say that the indivisibles are solid and for those who say they are planes; [for they can come to be neither more rarefied nor more condensed, since there is no void within the indivisibles.]

He has made the assumption that if the atoms had affections, either they would have one or they would have several, and has refuted both <disjuncts of the consequent> – since it is possible neither that they should have several nor that they should have one – so as to conclude, no doubt, that it is not admissible for them to have affections. He now makes the further point that this very thing follows, not only for those who posit indivisible solids, but also for those who <posit indivisible> planes, namely, its not being possible for them to have either one affection or several.

Next, in order to show this to be so, he makes the further point: 'for they can come to be neither more rarefied nor more condensed, since there is no void within the indivisibles'. For the followers of Democritus held that being rarefied or condensed resulted from the interweaving of the void, for the things that had a larger share of the void were more rarefied, those that had a smaller more dense. So if there is no void in the atoms, they will not be more rarefied or more dense. He does not urge this point on the grounds that if this is abolished all the other affections are abolished with it (for it is not the case that if they were neither more rarefied nor more dense they would not have any other affection either), but he has already refuted the general claim that they cannot have affections, and he makes this point by way of trying <the general thesis> out on some particular affections.

Those who talk of planes said that there is no void, and perhaps they too said that being more rarefied or more dense results from having a larger or smaller number of passages. What is more, he himself says elsewhere[163] about these <philosophers> that they 'do not even attempt to generate any affection'. So either this is what he means by 'For this consequence will follow in the same way both for those who say that the indivisibles are solid and for those who say they are planes' – I mean the inability to possess any affection – or that, if they have more than one affection, given that they are indivisible in potentiality, as has been said, it will follow that they both act and are affected at the same time in one and the same respect. For what we said in the case of the atomic solids, this we shall also <have to> say in the case of the planes.

The next remark 'for they can come to be neither more rarefied nor more condensed, since there is no void within the indivisibles' is to be understood in this way in terms of the second interpretation. For since he has used the proposition that several affections exist in the atoms in accordance with one and the same formula to refute the claim that the atoms have more than one affection as a whole at the same time it might have been said – and we ourselves have already raised this difficulty – that it is not necessary <for them> to exist in accordance with one and the same formula, but that different affections exist in different parts, it is for this reason that he adds 'for they can be neither more rarefied nor more condensed'. For if according to the other hypotheses being rarefied or dense results from the interweaving of the void, and there is no void in atoms, it is evident that they cannot be either more dense or more rarefied by comparison with each other in virtue of comparisons between different parts of themselves. So they will have one and the same formula and a single potentiality. For if the atom is homoeomerous with itself and lacking any distin- 175,1
guishing feature, obviously there will be a single formula for the composition of its parts and the potentiality of the whole of it will be

the same; so it will not be possible for it to be hot in one part but something else in another.

> **326a24** Again, it is absurd too that small things should be indivisible but not big things. [As it is, there is good reason why larger things should be broken up rather than small things; for the former – the large things, that is – disintegrate easily because they hit a great number of things; but why should total indivisibility belong to small things rather than to big things?]

Another argument. If, he says, atoms are indivisible by nature and not on account of their small size (for in that case they would not be atoms by nature; for those who do not posit atoms do not divide them to infinity, but halt the division on account of smallness, not, though, maintaining that they are indivisible by nature, but that they are divisible in potentiality to infinity) – so if the atoms are assumed to be indivisible by nature, it would be ridiculous, he says, to suppose that the small one were not divided but the big ones were. For why should indivisibility belong to the small ones but not to the big ones, and what formula would account for this? We do say – and with good reason – that the large bodies are easily divided by comparison with the small, since we suppose that all are by nature divisible to infinity (so, while they are <all> divisible, it comes about that the big ones are more divisible because they are more prone to collide with the objects which divide them, whereas the small ones escape the dividing object on account of their small size and slip away), but for those who say that they are by nature indivisible this is not possible. For natural indivisibility is just the same in what is large as in what is small.

> **326a29** Again, do these solids all have the same nature, or do some differ from others?

Another argument. It has to be the case, he says, either that they are all of one and the same nature, or that different <atoms> have different <natures>. If they all had one and the same nature, as, perhaps, do different drops of water or lumps of earth, what is the cause of their separation and segregation? Or again, why do they not coalesce when they touch, as when one drop of water touches another or one fire another fire? For being of one and the same nature, they ought to be able to coalesce with each other. Should it be said that they do not coalesce because of their hardness, in the same way that when a shell is broken it is no longer possible for the parts to stick together, first of all one needs to ask what the cause is of the separation and division, as has already been said, and then, if they are all of one nature in this way, they should not when combined

produce different natures; or, if they do, <different> shapes only, as in fact heaps of stones or shells produce, not <things of> a different nature, but only heaps of a different shape. But if they were of a 176,1
different nature, then clearly they would differ in quality and not just in shape. So, he says, it would not be they who would be the fundamental constituents[164] of things, but things which were more fundamental that they are, namely, the qualities in accordance with which these both exist and are divided.

> **326a31** ... – some, say, being fiery, some earthy in their bulk. [For if there is one nature for all of them, what is it that keeps them apart? Or why, when they touch, do they not come to be one, as water does when it comes in contact with water?]

i.e. if some had the bulk and corporeality of fire, some of earth.

> **326a34** For the later is no different from the earlier.

The later drop of water which comes upon the earlier and becomes one with it does not differ from it at all, and this is why it becomes one with it.

> **326a34** If, on the other hand, they are different, what sort of things are these? [Obviously these will have to be posited as the principles and causes of what result from them, rather than the shapes.]

There are two different readings here: 'How many are these?' and 'What sort of things are these?'[165] If 'How many are these?' were read, it would mean 'How many special features do their natures have?'; but if 'What sort of things are these?' is read, i.e. these things in accordance with which they differ, they would be qualities. So these would be the fundamental constituents of bodies, not the atoms.

> **326b1** Again, if they were different in nature, they would act upon and be affected by each other when they came in contact.[166]

Since it has been proved that wholly like things do not act upon and are not affected by each other, but <rather> things which are in one way alike and in another way unlike, if the atoms have different natures, it is absolutely necessary that they should be of this sort. For it is the case both that their substratum is one and the same, which is what makes them atoms, and that their qualities are different. Being like this, when in contact, they would both act upon and be affected by each other.

177,1 **326b2** Again, what is it that moves them? If it is something other than they, they will be capable of being affected after all. [If, on the other hand, each one moves itself, either it will be divisible into a part that moves and another part that is moved, or contraries will belong to it in the same respect, and its matter will be one, not only in number, but also in potentiality.]

Another argument. What, he says, is the cause of motion for the atoms? For they say that they are always in motion. If that which moves them were something other than them, they would no longer be incapable of being affected; for it is clear that it is in virtue of the fact that that <other thing> acts upon them and they are affected by it that they are moved. But if each moves itself, either it moves with one part <of itself> and is moved with another, or it both moves itself and is moved as a whole. For if the former, it will no longer be indivisible, nor even the same and homoeomerous (for it will be divided into mover and thing moved); but if the whole both moves and is moved, contraries will exist in the same thing, since the patient is the contrary of the agent. And its matter, he says, will be one, not only in number, but also in potentiality. If, he says, the same atom as a whole both moves and is moved, it will be one, not only in number, but also in potentiality. For what we maintain is that the matter is one in number, but in potentiality it is not one, but many; for it is <each of> the opposites in potentiality being able to change to either of them in turn. But if, though incapable of being affected, the same <atom> both moves and is moved, it will be at the same time one and the same thing to be able to move and to be able to be moved. Similarly also with the contraries. For the contraries are capable of acting and of being affected, and that which acts moves and that which is affected is moved. <So>[167] the same <atom> will have at the same time the potentialities of contraries, given that it is moved in respect of the very same thing in respect of which it moves. Therefore the potentiality will be the same in respect of which it acts and is affected, i.e. that in respect of which it heats and is heated, cools and is cooled. So the contraries will exist at the same time, which is impossible.

326b6 Others maintain that the affections come about through movement <through>[168] the passages. [Now if, on the one hand, <this happens> when the passages are full they are superfluous, for if in this manner the whole of a thing were affected in some way, it would be affected in the same way even if it had no passages and were itself continuous throughout.]

He has set out the ways in which his predecessors said acting and being affected took place, and, next, has indicated his preference for the theories of the followers of Democritus as conforming better to the empirical evidence and as capable of applying to all the things which nature brings into existence (and these are the theories which posit atoms and the void). Now, having made a start from this position on the job of refuting the atomists' hypothesis, he turns back finally to the object of the enterprise. And this was the refutation of the views of the earlier philosophers giving their account of how acting and being affected take place. They were the theory which maintains that this happens by means of passages, the leading proponent of which was Empedocles, and that which <maintains that it happens> by means of the void, as the followers of Democritus and Leucippus supposed. 178,1

So this is how he proves that acting and being affected do not take place by means of passages: We know that those who posit the passages do not suppose them to be empty but filled with a certain fine-textured body like air. In this way they differ from those who posit the void. So if the thing which acts on it comes into contact outside with the body surrounding the passages, e.g. the actual rim at the entrance to the passage, and acts in this way,[169] the passage <becomes> superfluous (for if its action had been produced entirely by contact outside, then even if <the thing affected> had not had passages, its action would have taken place by means of the contact).

But if the action is produced by its entering the passages, either it extrudes the body which is in the passages or it does not extrude it. If it does not extrude it, body will have passed through body; if it does extrude it, either it acts upon the bodies which border <it> on the inside and surround the passages, or it has no effect. If it has no effect, the passages contribute nothing to <the processes of> acting and being affected. If it does have some effect, does this come about by means of certain passages at this position as well, or simply by being in contact? If it happens by means of certain passages, we shall ask the same questions once more about them, and so on ad infinitum. So that in the end the whole thing will be passage. But if its action took effect on the parts within surrounding the passages, but simply by being in contact, obviously, even if there were no passages, by being in contact the one would act and the other be affected, and the passages would be of no use for acting and being affected.

In this way he demolishes the theory that acting and being affected take place by means of passages. He does not himself add that they do not take place by means of a void, whether because he has refuted the theories of those who maintain these things (it has been demonstrated in the *Physics*[170] that in general it is not possible for there to be a void; and he has shown both here and in other places that there are no atoms), or else – and this is more <to the point> – because whatever has been said on the subject of the passages can be appropriately said in precisely similar terms about the theory of the void also. For if the thing which acts does so by being in contact with the bodies outside which surround the void, the void is superfluous (for even if it did not exist the action would take place as a result of the contact), and if it happens in <the void>, the same things which were said about the passages can appropriately be said once again.

'Others <maintain that the affections come about> through movement ... the passages.' Either 'movement ... the passages' means 'the movement and entrance of the thing which acts through the passages', or <it means> 'in virtue of the fact that acting and being affected takes place first in the passages', i.e. in virtue of the fact that the body which is in the passages is affected and moved when the thing which acts has come in contact with it.

179,1 **326b10** Again, how can it come about in the way they say with regard to seeing through something? [There is no possible thoroughfare through the transparent <medium> either where there is contact or through the passages, if on the one hand, <this happens> when the passages are full.[171]]

He has said 'If, on the one hand, <this happens> when the passages are full', but has not balanced this with 'If, on the other hand, <this happens> when they are empty', that is, if <this happens> when the thing that acts extrudes the body which is in the passages. But we have spoken about this.[172] He himself reverts to showing that seeing does not take place either in the way it is said to do by these <philosophers>, by the visual rays making their way through the passages. For it would have to be the case that the visual rays either extruded the body that is in the passages of the transparent bodies and penetrating them in this way take hold of the objects of sight, or else without extruding anything passing through the body or through the contacts themselves, i.e. the interface of the two bodies, the body which is in the passage and that of which the passage is made, e.g. of glass.[173]

Manifestly, it is impossible for the visual rays, which on their own hypothesis are bodies, to pass through the contacts. For the boundaries, when they touch, coincide with one another and leave no room between, whilst it is not in the nature of the visual rays, being bodies, to pass through something which is not extended.[174] So it is not possible for the visual rays to pass through the contacts. And for this very reason it is not possible either for them to pass through the body that is in the passage; for body would be passing through body. Moreover, it would mean that the passages were of no use.[175] For if in general visual rays were of a nature to pass through body, they would also pass through the substance of the transparent objects and there would be no need of passages.

But if the visual rays cut the body which is in the passages into pieces[176] or even extruded it completely, and thus pass through and lay hold of the objects of sight, there are many absurdities implied by this account of the matter. First, we ought not to see anything all at once, but only bit by bit. For as each visual ray strikes against some part, since the visual rays do not form a continuum, it ought not to be possible for anything continuous to be seen.[177] Secondly, how do we see through the air? For if here too the air is cut <into pieces> and makes room for the visual rays, since we see a complete wall – And why do I say 'wall'? since we see at the same time the entire fourth part of the heavens – the space between, I mean that which stretches from the eyes to the thing seen, must needs be empty, which is impossible.[178] And since the actual object of sight is seen by more than one of us at the same time, it will follow that the visual rays, being bodies, will once again pass through each other, which is impossible. So from this it is obvious that bodies do not go out from the eyes.[179]

> **326b13** What will be the difference from having no passages? [The whole will be just as much a plenum.]

If, he says, it goes through the passages when they are full, why does it not also go through the transparent bodies themselves? For the full passages and the remaining body of the transparent things are on a par.

> **326b15** Furthermore if these, whilst empty, necessarily have bodies in them, this will again have the same result. [And if their size is such as to admit no body, it is ridiculous to think that what is small is empty, but not what is big, whatever its size, or to think that 'void' means something else besides 'place for a body', so that it is clear that for every body there will be a void of equal volume.]

If someone, he says, were to deny that there are passages through which acting and being affected or seeing took place, but said that these were voids filled with bodies, the same results would follow as in the case of passages. So how do they differ from passages? <It is> because the nature of the passages is never to be empty of body, whereas the void is both full <and not full>,[180] in so far as it is its nature to subsist *per se*. But if someone were to say that in the bodies which are in the void spaces there are certain passages in the middle, or that there are other void spaces, yet again we should find the same problems arising about *them*, whether or not the visual rays go through them although there are yet other bodies in them, and once more the same conclusions would be drawn.

But if it were said that the void spaces in the middle were too small to receive any body,[181] <the reply would be>, first, <that> the visual rays could not travel through them either, since they too are bodies (so the <void spaces> would be in vain), and secondly, he says, that, if in general it is the nature of the void to subsist *per se*, it is absurd and arbitrary to posit a void that is small, but not one that is large.[182] For what else is the void, he says, than space deprived of body? Then, whilst it would follow naturally to add 'So for every void space there will be a body of equal size', he himself says the converse of this: 'for every body there will be a void space of equal size'; for in this way he obtains the conclusion he wants. For if for every body there is a void space of equal size, and there is a body which is large, then there will also be a void space which is large.[183] But perhaps he did this because void space is convertible with body and body with void space. And for every body there will be a void space of equal size, but not a separated <void space>, as those <philosophers> posited, but one which has body in itself.

But it is not necessary for there to be body equal in size to every void space. For what if one were to suppose that there was a void space, and this was space deprived of body, but that nevertheless it never admitted body? So this position is not one we are forced to adopt, but it could, rather, be refuted on the grounds that it is arbitrary. For what is the absurdity in a void space being large, given that there is one which is small?

181,1 **326b21** Speaking generally, it is superfluous to posit the existence of passages. [If nothing can act by means of contact, it will not act by penetrating through passages. If, on the other hand, it can do so by contact, even if there were no passages, some things would be affected and others would act …]

He takes up the argument again and sets it out in a more precise way. So if, he says, it is not possible for something to act by touching from the outside, it will not act either by means of the passages; for it will merely enter the thing that is affected, but will not act in any way upon the bodies which line the passages, given that it was unable to act by touching. If, on other hand, it acted by entering and touching, it would act by means of contact even if there were no passages. And if someone were to say, 'But there is a need for passages if a thing is to be affected also in its innermost parts', it should be realised that this remark has nothing pertinent to say to the hypothesis. For it was not this which was in question: whether <it is true to say that> the thing that acts nowhere and in no way enters <the thing that is

affected> by means of passages. For it enters through the mouth and through the stomach and through the rest of the passages that are provided by nature, but the passages here do not contribute to <the actual process of> acting, but to the ability of the thing that acts to come close to the thing that is affected. The acting itself takes place by means of contact. It could be said that in this way passages are needed for the innermost parts to be affected, but the passages would be being said to be responsible, not for the action of the thing which acts, but only for the fact that the thing which acts and the thing which is affected touch one another.

> **326b24** ... provided they were of a nature to be related in this way to each other.

sc. <in terms of> acting and being affected.

> **326b24** From these considerations it is clear that speaking of passages in this way, as some have supposed necessary, is either mistaken or futile.

By 'in this way' he means there being passages in body already determined and existing in actuality, and being as it were themselves pierced with holes. But he does not refute the existence of passages *simpliciter* (for there are, as we have said, intestines, veins, arteries and the rest). Rather, as I have said, <he refutes the view> that the rest of the body is pierced with holes and that there are through the whole of the body passages that are imperceptible because of their small size. He says, therefore, that this sort of theory is, on the one hand, mistaken, since it posits things which do not exist, and, on the other, futile, because, even if we suppose them to exist, they will contribute nothing towards <enabling> them <to explain> acting and being affected.

> **326b26** Since bodies are divisible at any point, it is absurd to posit the existence of passages; for where they are divisible, they can be separated. 182,1

Even if passages were needed for acting and being affected, it was ridiculous, he says, to posit passages already determined and actually existing. For, since bodies are divisible at any point, that which acts, when it comes into contact and is about to act, can then make a passage for itself and divide the body. So there is no need for any passages already determined.

<1.9. ACTING AND BEING AFFECTED (3). ACTUALITY AND POTENTIALITY>

326b29 Let us accept, in order to say in what way generating and acting and being affected belong to the things that are, a principle that we have often mentioned.

After refuting the theories used by his predecessors to explain acting and being affected, he now at last intends to set out the correct view of how acting and being affected come about. And we must, he says, now once more set out as a basis for our explanation the principle which we have often spoken about, <the distinction between> the potential and the actual. For in truth, as he himself says in the *Physics*,[184] 'If this nature had been seen it would have enabled us to solve all the problems <that arise> in <the study of> physical things'. So since, he says, one thing is potentially such-and-such and another is actually such-and-such, e.g. water is potentially hot whilst fire is <so> actually, given that the thing which acts naturally wishes to make that which is affected like itself, when that which is capable of acting and which actually is <such-and-such> comes near and touches that which is potentially <such-and-such>, i.e. that which is capable of being affected, the potential <such-and-such> that is in it changes into the actual <such-and-such>.

It is absolutely necessary for the thing which is capable of acting to touch the thing which is capable of being affected either immediately or through some intermediary, when the intermediary is the sort of thing which can both act and be affected, to be affected by the thing which acts and itself to act upon the thing which is affected, as does the air; for in the baths we are not warmed immediately by the fire, but the fire warms the air and the air <warms> us. So there is no need of passages for acting and being affected, but only that these thing touch each other, unless someone were, he says, to mean by passages the parts of bodies which are more easily affected, as occurs with minerals. For there are certain veins in minerals stretching throughout them, as in the case of marble when, if it is white, <there are veins which are> black or different in colour, obviously <so> because they are of a different mixture and these parts are affected either more or less easily. So in this way, he says, it is not surprising if there actually are in things which are affected certain parts like this which stretch throughout and are more easily affected, and these 183,1 are affected more when the thing which acts comes into contact. So, he says, if someone thought it worthwhile to call these 'passages', let him do so.

There are those for whom this prompts the questions 'Why then are there not determinate passages anywhere or in any way in body?', 'What causes the daytime perspiration of bodies to take place without being noticed?', 'From where does sweat come through?',[185] 'How do we explain the existence of certain ointments that cause bodies to become rarefied or compacted?' In reply to this they say that so far as the rarefaction or compacting of bodies by certain ointments is concerned, we should not regard them as being rarefied, like leather or wool, by being stretched and having void spaces made in the middle of it, but as water when it is rarefied produces air, not by being cut into small pieces and taking empty space into the middle of itself, but by changing into a more refined substance; and this is how bodies are rarefied uniformly, becoming more refined throughout and as a whole. This then is clear.

And they say that the sweat and perspiration do not have to emerge through pores. For nature disperses excess and useless moisture by its power of secretion, not through certain passages, but continuously moistening one part <of the body> after another, and in this way excreting the moisture to the surface. For it is like, they say, a linen cord which has been made wet, where it is not merely that the substance of the water is deposited in the empty spaces, but also that the substance of cord itself has been moistened, not by means of certain passages (for it would result in the whole thing being a passage). This is how it should be thought to happen in the case of the body also. From the innermost parts in a continuous succession one part after another is moistened, and thus the moisture is separated out towards the surface.

However, while this account is plausible, it is in no way true. For if it were only a quality which was separated out from the innermost parts, what is said would be reasonable. If, however, it is not only a quality but some substance, like the drops of sweat or the clouds of vapour, one of two things would have to be the case: either body would pass through body, or it would be separated out through certain passages. And the same response will apply in the case of the linen cord; for if it is not through certain passages that the water penetrates into the substance of the cord, body must have passed through body. So either the actual substance of the cord is not moistened as a whole, but there is merely a juxtaposition of the moisture, or else, if it is moistened, the substance must be regarded as being affected in some way by the juxtaposition of the moisture and as undergoing a change of quality without the actual substance of the water entering into it; for this is impossible.

But maybe the moisture envelops completely the fine fibres without going into the interior of the substance, and because of the extremely fine-grained character of the juxtaposition it seems to have passed through the substance, as also in the case of the very fine-

184,1 textured powders the juxtaposition of different kinds of things practically escapes notice. Nevertheless, even the fact that perspiration is effected through passages requires that certain determinate passages should exist in bodies, but, as Aristotle says, 'since bodies are divisible at any point' we must suppose that during perspiration bodies are divided and after perspiration are knit together again. And it is not necessary that perspiration should take place simultaneously from every part <of the body>, in a way that would force us to regard it as being divided throughout, but <it comes> from different parts at different times.

And in what way generating and acting and being affected belong to the things that are. 'Generating' refers to change in respect of substance and 'acting' to change in respect of quality.

> **326b31** [If there is such a thing as what is potentially such-and-such as well as what is actually such-and-such,] it is natural for it to be affected, not in this place rather than in that, but everywhere, to the extent that it is such-and-such, [though to a lesser degree according to whether it is more or less such-and-such].

i.e. to the extent that it is potentially something, e.g. hot or cold; for to the extent that it is <such-and-such> potentially, it is natural for it to be affected as a whole. And, he says, that which is more or less <such-and-such> potentially is more or less capable of being affected; for that which is potentially hotter is heated more by that which is actually hot, but that which is less is not.

> **326b34** And in this way, rather, one could speak of passages, [as in minerals there stretch continuous veins of what is specially liable to be affected].

In this way, <i.e.> in virtue of the fact that some of the parts are capable of being affected to a greater extent that the rest, enabling one metaphorically to call the more easily affected of the parts passages. The fact that there are some parts in bodies more easily affected than others is indicated by the veins which stretch through minerals, often being more easily affected than the remainder, but perhaps also less easily affected *simpliciter*,[186] because there is a disparity in the parts of these things between being more and being less affected.

> **327a1** Anything which has grown together to make one thing is incapable of being affected. So too are things that are in contact

neither with each other nor with other things whose nature is to act and be affected: [I mean, for example, that fire heats not only when in contact with things but also when it is at a distance from them: for the fire heats the air and the air heats the body, air being of a nature both to act and to be affected.]

It is the nature, he says, neither of the parts of the continuous to act upon one another, nor of things which are altogether separated and distant <from each other>, but, things that act upon and are affected by <each other>, whilst being separated from each other, must be in contact either immediately or through certain intermediaries, the 185,1 intermediaries being such as to act and be affected naturally, being affected by that which is actually <such-and-such> and acting upon that which is potentially such-and-such. And he clearly sets out examples.

327a6 As for the view that things are affected in one part but not in another, in the light of the distinctions we made at the beginning there is this to be said. [For if the body is not everywhere divisible, but there are certain atoms possessed of size, whether solids or surfaces, none of the bodies would be everywhere capable of being affected, but nor would anything be continuous.[187] If, however, this is false and every body is divisible, it makes no difference whether it is divided into parts which are in contact or is divisible. For if it can be segregated along the lines of contact, as some say, even though it is not yet in a divided state, it will be in a divided state. For it is possible for it to be divided. For nothing impossible comes into being.]

Since he has said of that which is potentially such-and-such 'it is <a thing's> nature, not to be affected in one place but not in another, but <to be affected> everywhere to the extent that it is such-and-such', he now intends to say what sort of theory it is that has the consequence that the body is not everywhere capable of being affected, but in one place and not in another. For if, he says, the body is not everywhere divisible, but there are certain atoms possessed of size, whether solids or surfaces, none of the bodies (sc. of the composite <bodies>) would be everywhere capable of being affected, but nor would anything be continuous; for they will not be affected in their atomic parts, since these are incapable of being affected and indivisible. And this is why he adds 'nor would anything be continuous', because the indivisibles cannot be knit together (for if so they would also be divisible), but only lie alongside each other. So this theory implies that the bodies are not everywhere capable of being affected.

But if this theory is incorrect, and the true theory is that the bodies are everywhere divisible, it makes no difference, he says, whether we make the bodies actually divided but touching one another, as Plato said (for, <in his view>, the surfaces touch one another and acting and being affected takes place through the contacts),[188] or whether they are actually not divided, but potentially <so>; for that which is capable of being divided will be divided when the thing which divides it is present. So it is useless and superfluous for <the job of explaining> acting and being affected to posit in advance that the bodies are actually divided, but are merely touching each other along the divisions.

327a14 Generally, it is absurd to suppose that it is only in this fashion, as a result of bodies' being split, that anything comes to be. [This theory does away altogether with alteration; but we see the same body, remaining continuous, at one time liquid and at another solid, and this happens to it without division or composition taking place, or touching or turning, as Democritus says; for it has become solid from being liquid without any change of order or position in its nature, nor does it have within it the hard and solid bodies, indivisible in their bulk, but it is at one time liquid in the same way throughout, and at another time hard and solid. Again, there could be no such thing as growth or diminution; for it will not be true of every single part of it that it has become bigger, if something is added to it ...]

He has proved that, even if acting and being affected take place as a result of bodies' being divided, as Plato seems to maintain, saying that they take place by means of the contacts when the bodies are divided and touching[189] * * *. He now shows that, since body, being everywhere divisible, will be divided in the course of being affected and there is no need for it actually to have been divided before the affection <takes place>, it is, generally speaking, absurd to suppose in this way that acting and being affected take place as a result of
186,1 bodies' being split. For this theory, he says, does away altogether with alteration. First, because change of quality takes place while a body remains intact and in no way undergoes division, as in the case of the solidification of milk and water. For water or milk is frozen without being divided, and without undergoing, as <it would have to> according to the views of Democritus, *tropê* or *diathêgê*, i.e. changes in the order and position of their parts,[190] but also without the expulsion of any atoms (I mean those productive of moistness) and the retention of the hard ones. For before the water is frozen it is entirely homoeomerous, having no hard or indivisible parts in itself, but being wet as a whole in each of its parts. Similarly, when it is solidified, it is solid throughout.

Again, he says, there could be no such thing as growth or diminution. Those who hold this position, he says, do away, not only with alteration, but also with growth and diminution. For if it is true that every part of the growing thing becomes larger and <every part of> the thing which is diminishing smaller, it is impossible for this to be safeguarded on their hypothesis. For if it is by things' being added in the passages or in the void spaces or in the contacts that growth takes place, evidently the thing that grows has not grown at every point, as it would not either, if in a cistern water were added at some part: it would be said to have become more, but not to have grown.

327a24 ... rather than the whole having changed either through the admixture of something or because it changes in itself.

'Admixture', as in the case of nourishment (for the whole becomes larger by the admixture of nourishment which has been changed), 'or because it changes in itself', as when water changes into air; for in this case every part of the water becomes larger, but this is not growth, but coming-to-be. He says this indicating that they are not able to save[191] coming-to-be either. For if air <comes to be> from water, but comes to be as the result of the atoms of air which are in it being separated out, there is no necessity for a larger volume of air to come to be – as in fact it does. Either, therefore, these theories are correct and neither alteration nor coming-to-be will be saved, or if there is such a thing as alteration and coming-to-be, the theories are mistaken.

327a25 [This, then, is how we settle] that there are such things as generating and acting ... 187,1

If the words 'generating' and 'acting' are understood as applying to substance, it is the same thing as changing what is potentially substance to what is actually substance; but if 'generating' applies to substance and 'acting' to change in respect of quality, which changes what is potentially such-and-such to actuality * * *[192]

327a26 ... and coming-to-be and being affected ...

He <now> mentions the converse. Having spoken of the things capable of acting, he speaks also of the things capable of being affected. So 'coming-to-be' is being drawn from what is potentially substance into what is actually substance, and 'being affected' is being altered in respect of quality.

327a27 … and in what way this is possible …

Namely, by the matter being converted with respect to the whole of itself and changing.

327a28 … and in what way it is not possible though some say that it is.

For they say that coming-to-be and perishing take place by aggregation and segregation, and acting and being affected by means of passages; but he has shown that <this> is not possible.

<1.10. MIXING>

327a30 It remains to employ the same sort of method for the consideration of mixing, [since this was the third of the topics we originally set ourselves to examine.]

He said <earlier> that there were three things which needed to be understood and articulated from the discussion about the elements' changing into one another and about coming-to-be and perishing generally: What is contact? What are acting and being affected? and also, What is mixing? So having discussed contact and acting and being affected, he moves appropriately to the third of the proposed topics, to the discussion of mixing. And here he uses 'mixing' in the more special sense, to apply to mingling;[193] for 'mixing' is a more general term than 'mingling' and 'juxtaposition' and the rest, but mingling is also called 'mixing' in a more special sense, and it is that which is the topic of the present discussion.

There are, he says, five questions which need to be asked in connection with the theory of mixing, and he enumerates them clearly: what mixing is; and what it is that is capable of being mixed; and to which of the things that are it can be attributed; and how; and in addition to these he asks whether there is such a thing as mixing at all, which is the first of the problems.[194] For there might be those who doubt whether there is any such thing as mixing at all.

For it is bound to be the case, he says, that of the things that are mixed (since there must at least be two of them) either both are preserved in the mixing, or both perish, or one is preserved and the other perishes; there is no further possibility. So if it is proved that mixing cannot be preserved on any of these hypotheses, it looks as though it does not exist at all. For if both are preserved, there would be no mixing. For how would it differ from how it was before? For on this supposition they would be no more mixed before mixing than after mixing.[195] But if they both perish, they would not be mixed at all; for the things that are mixed must exist, so that they are in fact mixed, but things that have perished do not exist. And if one is preserved and the other perishes, this way too there would be no mixing. For, as I have said, the things that are mixed must exist, but in this case the one exists and the other does not. And moreover the things that are mixed have to be on a par,[196] but here they are not on a par, since one has perished.

These, then, and puzzles of this sort, are what do away with mixing. But he solves the puzzles by means of <the distinction between> the potential and the actual.

Since, he says, some things exist potentially and some actually, the things that are mixed might remain the same potentially in the mixing, but not actually. For in the wine that has been mingled,[197] both the wine and the water exist potentially, but I am not using 'potentially' in the sense I did earlier, in terms of suitability, in which water is potentially air, and not *simpliciter* in the second sense, merely in connection with a disposition, as is the sleeping geometer. Rather, in the way in which the geometer who is drunk and is trying to do geometry is actualised in terms of his disposition, but not purely[198] – that is how it is with the wine and the water in mixture;[199] for in the mixing each operates in a restrained way.[200] In this way both are preserved in potentiality, but in actuality neither is like it was originally; for their pure activity has been restrained and is no longer as it was before the mixing. So this is how we will find our way out of all the puzzles. For mixing occurs without both <ingredients> being preserved *simpliciter*, and without both perishing, and without one perishing and the other being preserved. The mixing is something else over and above these <three possibilities>; for <it happens> with both of <the ingredients> being preserved in potentiality, as has been said, but neither in actuality.

So it has been demonstrated that there is such a thing as mixing, since the puzzles which seemed to force us to contradict evident facts have been solved. For self-evidence agrees that there is such a thing as mixing, yet the puzzles that were brought to bear compelled us to raise doubts about what is evident. Now these have been solved, we can take it as given that there is such a thing as mixing.

'... the same sort of method ...' For his custom always is first to set out and refute the opinions of his predecessors, and it seemed, he says, to some of them that there was no such thing as mixing, and they also produced these puzzles. By 'in the same way' he means either that he is going through the puzzles to which we are to find the solutions, as he himself has said elsewhere,[201] or that he is asking the questions 'whether the thing exists', 'what it is', etc.

327a32 [We have to enquire what mixing is,] and what it is that is capable of being mixed[202] and to which of the things that are it can be attributed, [and how; and furthermore whether there is such a thing as mixing or whether this is false.

For it is impossible for one thing to be mixed with another, according to what some people say; for supposing that the things after being mixed still are and have not altered, they say that now they are no more mixed than they were before, but are just the same; and that if one of the two things is destroyed, they have not been mixed, but one exists and the other does not, whereas mixing is of things in the same condition; and that it is no different if, when the two things have come together, each of the things being mixed is destroyed, because they cannot be things that have been mixed if they cannot be said to be at all.]

Perhaps these remarks are pleonastic.[203] For it is that which can accept mixing which is capable of being mixed, and things like this are also the things to which mixing can be attributed. And 'how' means 'in what way are the things that are mixed mixed'. And last of all he places the first of the problems, if there is such a thing as mixing at all, but in his detailed exposition he demonstrates it first.

327b6 Now what this argument is after seems to be to clarify the distinction between mixing and coming-to-be or perishing, [and between a mixture and a thing that comes to be or perishes.]

The argument, he says, about whether mixing takes place with <the things that are mixed> being preserved or perishing seems to raise the question what the distinction is between coming-to-be and perishing and mixing. For if both or just one of them perished <in the process> it would obviously be perishing and coming-to-be <that occurred>, not mixing. So what is the difference between mixing and perishing and coming-to-be? For again, if <both> are preserved there cannot be mixing for the reasons given. So if mixing exists at all, he says, it will <have to> differ completely from coming-to-be and perishing.

327b9 So once these are clear the problems should find their solution.

By 'these' he means 'how mixing differs from coming-to-be and perishing'; for if <the answer to> this is found, the puzzles are solved.

327b10 On the other hand neither do we speak of the fuel as having been mixed with the fire …

After saying that there has to be a difference between coming-to-be and perishing and mixing, he wishes to prove this very thing, at the same time providing examples of what he had said earlier, that mixing involves neither both <the ingredients> perishing, nor <just> one of them, nor both being preserved. But the connecting phrase 'On the other hand'[204] obscures the continuity with what has gone before. So he should have spoken in this way: 'for we do not speak of the fuel as having been mixed with the fire'. For wood is not mixed with fire, but what happens is perishing of wood and coming-to-be of fire. And this would be an example of one thing being preserved and another perishing.

327b12 [… nor, when it is being burnt, that it is being mixed,] either with its own parts or with the fire: [we say rather that the fire comes to be and that the wood is destroyed.]

This is an example of the case where both perish. For neither would the parts of the fuel for the fire be said to be mixed with each other in the fire on the grounds that just one flame results from them. For as I have said, what occurs is perishing of the wood and coming-to-be of the fire, but in no way mixing. And by the 'parts' of the fuel for the fire he means the different pieces of wood. So the pieces of wood, out of which the fire comes to be, are not mixed with each other in the flame, but perish. And 'or with the fire' should be understood in the sense of 'or even with the fire'. For he said already 'neither[205] do we speak of the fuel as having been mixed with the fire', so that it is superfluous to say the same thing again, unless, as I have suggested, it is understood as meaning that the pieces of wood are not mixed with each other just as <they are> not <mixed> either with the fire.

327b13 Similarly we do not say that the nourishment is mixed with the body …

This too is an example of one thing perishing and another being preserved. But perhaps it is because he wishes to distinguish mixing also from growth, as well as from coming-to-be, that he uses this example too.

327b14 … nor the shape with the wax [when it impresses itself on the lump. Nor can body and white, nor in general can affections and dispositions, have been mixed with the things that have them; for we see that they are preserved.]

This is an example of both being preserved. For both the wax and the shape are preserved, but one would not say that the shape is mixed with the wax; for one is substratum and the other is form, and form and substratum do not belong to the category of things that are mixed. So absolutely none of the things that are *in* a substratum is said to be mixed, neither the dispositions nor the affections, whether mental or corporeal. And by these means he could have distinguished mixing from alteration.

327b17 No more can whiteness and knowledge have been mixed, nor can anything else of the non-separables. [This is indeed the unwelcome consequence of saying, as some do, that everything at one time was together and in a state of having been mixed. For not everything is capable of being mixed with everything else. When two things are mixed each must exist as a separable thing, and no affection is separable.]

This too is an example of both being preserved. For when whiteness and knowledge exist in one and the same animal they could not be said to be mixed; for they are both preserved unaffected by one
191,1 another. And he adds this example too in order to show that these things, which neither act upon nor are affected by one another, nor are separable, could not be mixed, even if they existed in one and the same substratum. You could take <an example of being> in one and the same substratum in a stricter sense,[206] in which two affections, sweetness and yellowness, are existent in honey. These would not be said to be mixed, since both are preserved and are not separable or able to subsist in their own right; for things that are mixed must be of that sort.

None of the things whose nature it is not to subsist in their own right could be said to be mixed, and it is for this reason that quality is not mixed with substratum (for quality is not separable), not form with matter; for neither is separable. You have here one characteristic property of things which can be mixed, or to which mixing can be attributed, its being necessary for them to be able to subsist in their own right; and he continues by setting out fully their characteristics.

327b22 Since, however, some things that exist, exist potentially and some actually, [it is possible for things that are mixed to exist in a sense and not to exist. Some other thing which comes to be from them exists actually, while each of the things which existed, before they were mixed, still exists, but potentially, and has not been destroyed. This is the solution to the problem raised by the previous argument.]

The solution of the puzzles begins here. Since, therefore, he says, some things exist potentially and some actually, it is possible for things that are mixed to exist in a sense and not to exist; for that which results <from the mixing> is potentially each of the things out of which it has been mixed, but actually neither of them. So the puzzles are solved, since the things which are mixed are preserved in one sense, and not in another. For this was what the puzzles meant: that they were preserved and not preserved.

327b27 Moreover, things that are mixed manifestly come together from having formerly been separate, and are capable of being separated again. [So neither do they both remain in actuality like the body and its whiteness, nor do they perish – either of them or both – because their potentiality is preserved.]

Mixing, he says, can be attributed to things which both subsist in advance in their own right and are naturally capable of being separated again after they have been mixed. And this, not because the things that have been mixed exist in the object in actuality, so that they can again be separated in the same way (for they would not have *been* mixed if they still existed in actuality and without any change), but he says this because the things which are mingled, since they do not lose their complete form but only the pure version of it, are naturally capable by means of certain separating and altering instruments of regaining once more their own form whole and entire.

They say indeed that mixed wine separates into wine and water when strained through what are called in common speech 'river soldiers';[207] for only the water is drawn, the more so when the wine is thicker. So either this is what he means by 'capable of being separated again', or, if this is thought to be impossible, since probably things 192,1
that have been mingled could hardly literally have been separated again, we should understand 'capable of being separated' as meaning 'being of a nature to subsist again in their own right in so far as in themselves they are capable of this', if, that is, there should be something with the power to segregate them.

This then, as we have already said, is one characteristic property of things to which mixing can be attributed, which, he says, do not remain in actuality in the <process of> mixing, as do the body and the whiteness that is in it, and do not perish either, whether just one <of them> or both. And this is how mixing differs from both coming-to-be and alteration. For neither is the form separable from the matter, nor do the ingredients in mixture exist in potentiality in the way in which water is said to be air in potentiality. For in the case of coming-to-be the matter of the air is air in potentiality but water in actuality, whereas in mixture the ingredients exist in it in potentiality – not their matter, but the forms themselves in a restrained state; and this is why <they exist> potentially, on account of not being pure and not being like they were before the mingling.

In making these points he is attacking the school of Anaxagoras, which said that at one time 'all things were together'; for unless everything was mixed with everything (for affection is not mixed with affection), how should all things be together?[208]

327b31 For this reason we can dismiss these things.

sc. the business of the puzzles.

327b31 The problem following on from this must be divided: [whether mixing is something relative to perception. For (i) when things that are being mixed are divided up so small and the pieces placed side by side in such a way that each of them is not apparent to perception, have they then been mixed?]

The word 'divided' is equivalent to 'examined', or else it is like saying that the problem itself must be treated by means of division.[209] For he now immediately investigates what mixing in fact is, and pursues the argument by means of division. For he proposes two ways of mixing, one relative to perception, where the ingredients are broken up into small pieces and the fragments juxtaposed to each other in such a way that there is no awareness of the juxtaposition because of the smallness of the fragments, but there seems to be mixing and mingling, as <would happen> if someone were to mix fine flour <made> of wheat with barley meal. For there is juxtaposition involving these, but because perception cannot cope with the smallness of the fragments, there will seem to be a mingling and the whole <mass> would appear to be homoeomerous.[210]

The same thing occurs in the case of powders, e.g. the one which is composed of six ingredients.[211] So this is one way of mixing which he proposes – one in which the forms of the ingredients are preserved, but it passes unobserved because of the minute size of the things juxtaposed. The second way, however, does not involve the preservation of the forms of the ingredients, but their being broken down into the things of which they themselves are composed, and their elements in this way being juxtaposed <to each other>, as maintained by the proponents of the atoms. For after fire and water have been broken down into the atoms of which they are composed, and after these <atoms> have in turn been placed alongside each other in such a way as to juxtapose each atom of fire to an atom of water, the mingling is produced in this way, as would happen if a theatre and a house were demolished and the stones with which, perhaps, the theatre and the house had been built were placed alongside each other.

After refuting <the adequacy of> each of these proposed ways of mixing, he thus proposes in his own person the true one. And in order to refute them he uses two assumptions, one that mixing is different from putting together, and the other that every body and, in general, everything that is continuous, is divisible ad infinitum – and of these he has demonstrated in the *Physics*[212] the thesis that that which is continuous is divisible ad infinitum, and he accepts the thesis that mixing is different from juxtaposition as a common assumption agreed by everyone. For no one says in the case of mixture that wine and water have been put together, but that they have been mixed or mingled, not that a house is a mingling of stones and wood, etc., but that it is a putting together, taking for granted that we know that mixing and mingling are different from putting together. With both of the assumptions he refutes both of the theses:[213] using that which says that everything continuous is divisible ad infinitum, he refutes one thesis <and says> that mixing does not have[214] the character attributed to it by those who maintain that mixing is achieved by breaking down <the ingredients> into more elementary <parts>, i.e. atoms, and juxtaposing these each one to each one – the view of the school of Democritus. For if body is divisible to infinity, it is impossible for each part to be put alongside each part; for it will always be the case that the parts lying alongside each other, being divisible, will not have each <of their> parts lying alongside each. For if a part is divided into three, the middle segment is pure without anything of a different nature lying alongside it. So if the division were into atoms, once the ingredients had been broken down into their elements it would be possible for every part to be juxtaposed to every other part. Since, however, whatever is continuous is divisible ad infinitum, it is accordingly impossible for every part to be juxtaposed to every other part.

Moreover, if it has been demonstrated that mixing does not involve the destruction of the ingredients, it is accordingly impossible that mixing should take place by means of things' being broken down into atoms; for in that case fire and water would perish if they were mixed. For it is not each spherical atom that is fire, but the piling up of these <atoms>. And the same holds for water. So if the ingredients were broken up into more elementary <components>, they would be destroyed; but mixing is of things which exist, not of things which perish.

Thus one of the assumptions has refuted one of the theses and the other assumption equally refutes the other.[215] For if mixing is different from putting together, mixing would not be something that was relative to perception. For if it were a matter of small things' being juxtaposed to each other, so that their smallness resulted in the putting together being outside the scope of perception, it would be putting together and not mixing. For it would be the case, not that the fact of the juxtaposed items' not being seen because of their smallness made it mixing, but that their being juxtaposed rather than mingled, even if not apparent to sense, meant that it was not mixing.

What is more, if mixing were relative to perception, the same thing would be mixed for one person, who lacked keen eyesight, and not mixed for another person, Lynceus for instance, who is said to see beneath the earth because of the keenness of his vision. This would mean that the being of mixing would depend on a relation, not in <its own> nature – and that is illogical; for things that are mixed, even if there is no one to discern <the fact>, are manifestly things that are mixed.

> **328a1** [Or (ii) is it not so,] but <the situation is> such that every single part of the ingredients is alongside every single part.

The second thesis. Obviously this involved breaking the ingredients down into more elementary <components>, i.e. atoms, and juxtaposing each one to each one; for otherwise it is impossible.

> **328a2** So that is the way in which it is said, for example, that grains of barley are mixed with grains of wheat, [when every <grain> of each is placed alongside every <grain> of the other.]

He provides an example of the way <of speaking>, the way in which mixing is said to be relative to perception. But he is thought to be making these remarks in connection with the second way <of speaking>, since he goes on to say 'when every grain of each is placed alongside every grain of the other'; for he uses the same form of words as he used in connection with the second way <of speaking>. But this is a mistake: he is saying this with regard to the first way. And this is clear from the fact that he says '*that* is the way in which it is said', obviously meaning the previous one, not the one that is coming. And he says 'every <grain> of each alongside every <grain> of the other', i.e. barley alongside wheat,[216] intending us to understand that each grain of barley is alongside some one[217] grain of wheat, rather than in heaps – many grains of barley being together with many grains of wheat. But the example would fit better the second way <of speak-
195,1 ing>; for in the case of things which are hard to perceive because of their minuteness (e.g. grains of powder, mixture of barley meal and wheat, and some other cases) absolutely every one of the ingredients does not lie alongside absolutely every other, but it is possible for

some particles of the same sort, e.g. barley or some other kind, to lie alongside <each other>.[218]

> **328a3** But if every body is divisible, given that a body is indeed capable of being mixed with a body, each part would have to come to be homoeomerous in comparison with each part. [But since there is no such thing as a thing's being divided into parts which are the smallest possible, and since putting together is not the same thing as mixing but different, it is right, clearly, to say that things have been mixed neither (i) when the things being mixed are preserved at the level of small particles (for this will be putting together and not mingling or mixing, nor will the part have the same proportion as the whole; but we say that if things have in fact been mixed the mixture has to be homoeomerous, and that just as a part of water is water so it is with what has been mingled. If, however, mixing is just composition at the level of small particles none of these consequences is maintained, but 'being mixed' will be relative to perception: one and the same thing will be mixed for one man whose sight is not sharp, whereas for Lynceus nothing is mixed)] …

At this point he begins to refute arguments for the two ways of mixing that have been described. And he uses a hypothetical syllogism like this: If every body is capable of being mixed with a body, and if body is divisible ad infinitum, it follows that each part of a body that is mixed is homoeomerous. For 'Each would have to come to be homoeomerous in comparison with each' is equivalent to 'It has to be the case that every part will be homoeomerous with every <part>.' This is what the hypothetical is like, and its antecedent, as I have said, is <the conjunction of> two assumptions, <namely,> that body is divisible ad infinitum and that mixing is different from putting together.

For 'mixing is clearly different from putting together' is potentially contained in 'given that a body is capable of being mixed with a body'; for in putting together or juxtaposition each is not juxtaposed with each, since body is divisible ad infinitum.

And if every part of the mixture must necessarily be homoeomerous, it would be the case neither that mixing was the putting together of small pieces, nor that each was alongside each in terms of breaking <the thing> down into atoms; for the atoms lying alongside each other are not like one another, and it is an assumption that body is divisible ad infinitum. Where <do we get this> from? From the fact that the consequent follows from the antecedent, that if mixing is different from putting together, and putting together involves anhomoeomerous parts, mixing should not involve anhomoeomerous parts. So <the parts are> homoeomerous. But not homoeomerous in the way that those who speak of atoms would have it. For since they maintain that composite things are mixed together out of atoms (for the atoms themselves are not <mixed>), and they are mixed by being broken down into atoms and by the juxtaposition of each atom alongside each, they would call the whole homoeomerous, meaning by 'parts', not the atoms (for these are <rather> elements), but the parts of the whole which are composed of them, in all of which the juxtaposition is seen to be alike.[219] It is evident that what is mixed is not homoeomerous in this sense from the fact that it was accepted in advance that every body is divisible ad infinitum, but that as long as the division goes on (i.e. to infinity) all the parts are found to be homoeomerous.

So this, as I have said, is what the hypothetical is like. The minor
196,1 premiss is potentially the same as the antecedent in the hypothetical, but the expression of it is different. For instead of giving as the minor premiss 'But every body is divisible and mixing is different from putting together', he uses a potentially equivalent premiss 'Since there is no such thing as a thing's being divided into parts which are the smallest possible, and since putting together is not the same thing as mixing', and of this 'since there is no such thing as a thing's being divided into parts which are the smallest possible' is equivalent to 'since every body is divisible', and 'nor is putting together the same as mixing' which is contained potentially by the second conjunct of the antecedent, is equivalent to 'given that a body is capable of being mixed with a body'.

When these have been taken as minor premisses, the conclusion follows that what is mixed is in every way homoeomerous with itself, which is not itself posited, but is itself argued for by other <propositions>. 'It is right, clearly,' he says, 'to say that things have been mixed neither when the things being mixed are preserved at the level of small particles; for this will be putting together and not mingling or mixing, nor will the part have the same definition as the whole.'[220] These remarks are designed to prove that what is mixed is homoeomerous; for its being homoeomerous is inferred by refuting <the claim that it is> anhomoeomerous. 'But if mixing is putting together at the level of small particles, none of these consequences will hold;' for the whole will not be homoeomerous. 'Nor will the part have the same definition as the whole,' which followed the assumption that putting together is different from mixing. But if the whole is not homoeomerous, mingling is merely apparent and not real, and for this reason it would not be mixed for Lynceus.

328a15 ... nor (ii) when things have been divided [so that every part of one is alongside some part of the other, since it is impossible for things to be divided in this way.]

'It is not mixed' is shared <with the 'neither' clause>.[221] He turns to the other way, that which says that mixing is each <part> being alongside each <part>, the occurrence of which is itself too a consequence of putting together at the level of small particles. For given that body is divisible ad infinitum, it is impossible that each <part> should be alongside each <part>; for it is impossible for what is continuous to be divided into atoms. So the occurrence of this will be a consequence of putting together at the level of small particles as well, which is absurd in addition to the fact, which I have already mentioned, that the nature of the ingredients is not preserved, as it should be, in the process of breaking them down into their elements.

> **328a17** So either there is no such thing as mixing or we must again try to say how it takes place.

The overall structure of the argument was 'If there is such a thing as mixing, the parts of what is mixed must be homoeomerous'. So two ways of mixing have been proposed and each has been refuted by showing that neither is able to preserve <the truth that> the parts of things that have been mixed are homoeomerous. So once this has been established, he says, it will follow either that there is no such thing at all as mixing, or that there is some other way of mixing besides those which have been mentioned. And he adds 'again' because he has done the same thing as he did in the case of the earlier things <he has discussed>:[222] rejecting the views that have been refuted previously, he himself shows how each of them takes place.

> **328a18** There are, as we said, some beings which are capable of acting and some which are capable of being affected by these others. [Now for some things there is conversion, those, namely, which share the same matter, being able both to act upon one another and to be affected by one another. Some things, on the other hand, though they act, are incapable of being affected, those, namely, which do not have the same matter.]

Having rejected the false notions about mixing, and wishing finally to say what mixing is and how it comes about, he reminds us of the things that have been said about acting and being affected, namely, that, of the things which act, it is those which share a common matter with each other which are also mutually affected, and those which do not have a common matter which are not mutually affected. So since it is the things which act upon and are affected by each other that are mixed, and things which are alike in matter act upon and are affected by each other, it is these things, which share the same matter, which will be mixed with each other. So this is what mixing is: the action upon one another of things alike in matter.

> **328a22** So of these there is no mixing. And that is why medical skill does not produce health, nor health by being mixed with the bodies.

There is no mixing, he says, of these things – namely those which do not share the same matter and for this reason are not mutually affected either in the course of acting. And that is why, he says, medical skill is not mixed with the body when it acts upon it; for medical skill and the body do not share the same matter, because instances of knowledge are in the soul. So for this reason medical skill when it heals the body is not mixed with it, given that those things are mixed which are mutually affected by each other, and it is things which share the same matter which are mutually affected.

But there is some difficulty in the reading here. 'And that is why,' he is saying, 'medical skill does not produce health by being mixed, nor again does health produce health by being mixed with bodies'. So either 'And that is why medical skill does not produce health [by being mixed with bodies],[223] nor does health by being mixed[224] with bodies' is to be understood in this way, where 'being mixed' is used only once to apply to both subjects.[225] But clearly health, being a form, is not mixed – he has said that forms are not mixed with their subjects. Or we should take the health that produces to be the formula <of health>,[226] and that which comes to be the health that is in the body. Neither of these is mixed with the body. For the one has its being in the soul, and the other is a form of the substratum.

> **328a23** But amongst things which are capable of acting and of being affected, those which can <easily> be divided, when many of them are juxtaposed to few or large ones to small, they do not give rise to mixing, [but to growth on the part of that which is dominant; for the other changes into the dominant one. Thus a drop of water is not mixed with ten thousand pitchersful of water, for its form dissolves and it changes into the totality of water. But when the two are more or less equal in strength, then each changes from its own nature in the direction of the dominant one, though it does not become the other but something in between and common to both.]

Having said what mixing is and what it is of, namely, things that share the same matter, since not all things that share the same matter are mixed (for stone is not capable of being mixed with wood), he therefore proceeds to determine what sort of things of those which share the same matter are the ones that are mixed, <and to state> that they are those which are easily divided. And by 'easily divided' he means those which are easily bounded, and it is the liquid[227] things which are easily bounded. So of the things that share the same matter, those which are liquid are the ones which are mixed. And he proceeds to explain the way in which liquids are mixed, namely, that it is not in whatever way they happen to be; for it is not the case that if any liquid you like is introduced into any liquid you like, they produce a mixture. For it is not the case that if a drop of wine, as he says, is put into ten thousand pitchersful of water, it makes the whole something mixed; for the fact that the power of the water overwhelms it causes the drop of wine merely to be affected by it, and it is no longer strong enough also to act upon it in return, but the drop of wine merely being affected changes its form, and what takes place is perishing and coming-to-be, not mixing. For though some things are of a nature to act upon each other, still they do not do so in any condition whatsoever, but <only> when they also have a certain proportion in respect of quantity; for forms need a certain quantity in order to subsist, as he says himself in the *Physics*.[228]

So even if there is in a drop the form of wine, nevertheless it is not in its power to affect ten thousand pitchersful of water. And from the start it was clear that mixing takes place in the part that comes near, and when the drop is diffused over a large <area>[229] and the quality is continually blunted[230] by the fact that there is more of that with which it is mixed, eventually it actually changes altogether. So the ingredients, he says, must in some way be balanced in respect of their powers. And he does well to say 'in some way'; for they need not be equal in every way, but the one must not exceed the other by too much. So if the powers are in some way balanced each will overpower the other to a certain extent, not however so as to destroy it altogether, but merely so as to restrain its excesses;[231] and thus when the two have been mixed the whole comes to be something common and not just one of them, as in the case of the drop put in <the large quantity of water>. So that which has come to be is common to them both, but each is preserved potentially in the mixture in the way that has been described.

> **328a31** So it is clear from this[232] that of agents, those are capable of being mixed which have a contrariety [(for it is these which are capable of being acted upon by one another)]. 199,1

It is clear, he says, from what has been said what are the things which can be mixed, namely, that they are those <members of the class> of agents and patients which have a contrariety. He says this because of celestial things; for these too act upon things here,[233] but they do not have a contrariety, because they do not have a common matter either.[234] So those <members of the class of> agents and patients which have a contrariety, if they are liquids and in some way balanced in their powers – these are the things which act and are affected.[235]

> **328a33** Small quantities, too, put alongside small quantities mix better, [because they change one another more easily and quickly, whilst large quantities acted upon by large quantities take a long time to do this.]

Having shown that mixing is neither a matter of juxtaposition nor the result of division into small quantities, he now says that even if mixing is not juxtaposition, it is still a help towards mixing if things are divided into small quantities and set alongside each other; for ingredients are mixed more readily when they are divided into small quantities. For what is small, he says, will be affected more easily by what is small than large things will be by large things, and what there is much of, he says, takes longer to be affected by what there is much of. And this seems in some way to go beyond the analogy. For why is it not the case that just as the ratio of part to part is the same, so also is the ratio of whole to whole; and so by parity of reasoning the whole will be affected in the same way as the part by the part?[236] But if this were so, the division into small quantities and juxtaposition would contribute nothing to increasing the speed of the mixing. However, even if it is not divided, but the whole is juxtaposed to the whole, they will none the less be mixed when they are affected by each other.

Moreover, if the more the powers are collected together, the more intense they are and the more they exceed the proportion of the parts, as is shown clearly in the case of weights (for when heavy things are divided they do not preserve the proportion of the whole – and this is most evident in the case of gold – but the parts are observed to be deficient in some degree in comparison with the weight of the whole. Similarly, when the parts are reassembled the whole is observed to be heavier than the ratio of the parts <would require>.[237] And so with fire, the more it is collected together, the greater its heat, which is why a large chunk of wood thrown into a large furnace ignites all at once, but in a smaller flame a smaller piece of wood would not be affected proportionately); if, therefore, the more the powers are collected together, the more they are intense, it follows that they would be more mixed if they were not divided.

In reply to this, we say that, as far as the example of the fire and the wood is concerned, it is only to be expected that the more fire there
200,1 is the more effect it has, because it is a case of coming-to-be and perishing. So, given that the fire alone is acting and the wood alone

being affected, it is only to be expected that the more the active power is collected together the more effect it has, either because it is not reciprocally affected at all by that on which it in turn acts, or because <it is affected> very faintly, and the more so since it is also growing; for the affair is one of coming-to-be and perishing. As far as the case of the weights is concerned, this is a meeting, not between opposites, but between like and like, and when like approaches like it coalesces and vivifies the whole.[238] And for this reason division makes the thing that is divided more susceptible to destruction, as was said above; so it is only to be expected that unification is more conducive to life.

So much for these cases. In the case of the things that are mixed, however, since the ingredients have a reciprocal and equivalent effect on each other, and act upon each other by being in contact with one another, and are more in contact when they are divided (for things that are massed together have great depth), it is accordingly only to be expected that when they are divided they will mix better, because they also act and are affected to a greater extent. And this is because they are more in contact. For even in the case of the fire, the same amount of wood would be more quickly affected if it had been divided, since its parts would have more connection with the fire. So in the case of the things that are mixed, since both must needs be affected, both have to be divided if they are to be affected more quickly; for what is affected is affected more quickly if it is divided. So it is well said that the ingredients mix better if they are divided and placed alongside each other in small quantities.

> **328a35** Accordingly amongst things which are divisible and capable of being affected those which are easily bounded are capable of being mixed, [since they divide easily into small parts ...]

What he previously called 'divisible' he now calls 'easily bounded', so it is clear that it was the easily bounded things that he was calling 'divisible' earlier.[239] So of things that are divisible those that are easily bounded (and these are the liquids; for a stone is divisible, but not easily bounded) – these are capable of being mixed, because they are easily divisible, and we have been saying that things that are divided mix better.

> **328b2** ... which is precisely what it is to be easily bounded.

This, he says, is the definition of 'easily bounded', being easily divisible. For that which is readily marked off from its surroundings is easily bounded, and this is the sort of thing that the easily divisible thing is, e.g. a liquid; for what is hard is not easily bounded.

> **328b3** [For instance, liquids are the type of body most liable to mixing.] For liquids are the most easily bounded of divisible things, unless they are viscous [(these have the effect only of multiplying and increasing bulk)].

He adds another further restriction, namely, that not every liquid is
201,1 capable of being mixed, but <only> that which is not viscous. Viscous things include olive oil, bird-lime, liquid pitch and everything of this sort; for owing to their viscousness and elasticity these things are hard to divide and hard to separate, and when things are poured over them they slip around them and do not sink to the bottom, and for this reason they do not mix either. For they are less easily bounded than things which are not viscous, and on this account when they come together they totally refuse to mix but merely increase the overall bulk.

> **328b6** When only one <of the things to be mixed> is capable of being acted upon, or where one is extremely likely to be and the other very slightly indeed,[240] [the mixture <which results> from both is either not at all, or only a little, greater in quantity. This is what happens in the case of tin and bronze.]

When, he says, only one is capable of being affected by the other, which is not itself capable of being affected by *it*, or <when> one is affected strongly and easily, the other not easily and <only> slightly – in these cases he says that the two together do not increase the bulk of either of them in the mixing, or, if there should be any addition, this is small, as though the greater <part of it> were used up, which, he says, is what happens in the case of tin and bronze; for the tin, being airy and full of air and liable to change into this <element>, completely evaporates and is volatilised when heated by the fire in the process of being mixed with the bronze, and merely tinges the bronze, but does not add anything to its quantity as well. Maybe, though, there is no mixing at all of the tin with the bronze, but <only> a juxtaposition and adhesion <of the one to the other>. For even when the bronze is liquid it is not mixed with it, nor does it penetrate through the whole of its substance, but only tinges the surface. So the tin resembles rather some form in the bronze as a substratum, as he says himself as the text proceeds.

> **328b8** For certain of the things which are stammer and are ambiguous in relation to one another: [they have the appearance, as it were, both of being a faint mixture and of being related as form to that which receives form. And this is precisely

> what occurs in this case; for the tin, like some affection of the bronze existing without matter, vanishes and once mixed departs, having effected nothing but a change of colour. This same thing occurs also in other instances.]

For stammerers both talk in some way and do not talk owing to the inarticulate manner in which they produce their words, and certain phenomena of this sort, he says, occur too where things are concerned, things whose mixing is blurred and unclear and lacking in articulation, which in one way are mixed and in another way are not. And having said that certain of the things which are stammer and are ambiguous in relation to one another, he explains this by adding they have the appearance, as it were, both of being a faint mixture and of being related as form to that which receives form. So the tin is in one way a form and affection of the bronze, evaporating in the mixing process and merely colouring the bronze, whereas <it was agreed that> affections were not capable of being mixed with their substrata; however, the tin is not altogether like this. This is why he also says that it 'stammers' and 'is ambiguous'; for a certain slight growth of the substratum occurs, and this does not happen in the case of forms. And the mixing that takes place in the case of ash and water is like this. For with these too almost all that happens is that the water is coloured by the ash and grows slightly, because a certain growth of the ash occurs in the water.

> **328b14** From what has been said, therefore, it is clear that there is such a thing as mixing, [what it is, and what causes it, and what sorts of things are capable of being mixed, since they are certain things with such a character that they can be acted upon by each other and can be easily bounded and easily divided.]

i.e. it exists in reality, and not, as some used to think, only in appearance, but not really. 'And what it is.' He himself next defines what mixing is, that it is 'the uniting of possible ingredients after they have been altered', these being capable of being affected and of acting in respect of quality, neither remaining the same in actuality, but both <doing so> in potentiality. 'And what causes it', <namely> because a common matter underlies the ingredients and because they are capable of acting upon and being affected by each other on equal terms. 'And what sort of things are capable of being mixed'. For neither the solids nor all the liquids <are so capable>, but only the things which are easily divided and easily bounded, provided they are not viscous or liable to slip around <each other> in the process of division.

328b17 For it is not necessarily the case that these things perish after they have been mixed, nor that they are *simpliciter* still the same …

For mixing is not destruction, but both <the ingredients> remain potentially what they were before they were mixed, not actually though, but having their peculiar qualities restrained,[241] but each of them needing an addition in respect of its own proper quality for it to come to be in actuality what it was, not however a total change; for this sort of change is coming-to-be and perishing and not mixing.

328b19 … nor that their mixing is a case of *composition*, nor that it is merely relevant to perception. [Rather, that is capable of being mixed which, being easily bounded, would be[242] capable both of being affected and of acting, and that with which it is mixed is of the same sort …]

* * * [243] … who said that any part whatsoever is placed alongside any part.

203,1 **328b21** … for what is mixed is relative to something homonymous …

Having said that 'that is capable of being mixed which, being easily bounded, would be[244] capable both of being affected and of acting, and of mixing with what is of the same sort as itself,' he adds 'for what is mixed is relative to something homonymous', using 'homonymous' in place of 'synonymous', but the usage is habitual with him.[245] For if both are easily bounded and capable of being affected by each other, they are obviously synonymous in this respect. For they are both capable of being mixed and conform to the definition of 'capable of being mixed'.

328b22 … and mixing is the union of the ingredients after they have been altered.

i.e. union by means of alteration. For when the ingredients have been altered in such a way that what emerges from them comes to be some single homoeomerous whole, then they are united. And this could be the definition of mixing: 'union of the ingredients by means of alteration', so that in one way the ingredients are preserved and in another way they are destroyed, not being in actuality such <as they were>, as he said, but being restrained[246] they seem to have perished, no longer preserving the full strength of the appropriate form.[247] However, in virtue of the fact that their qualities have not changed in respect of the whole of themselves, but that rather a union of these has remained, in this sense they remain and are preserved.

Lecture Notes taken by John Philoponus the Scholastic of Alexandria from the Course given by Ammonius the Son of Hermeias with some Observations of his own on the Second Book of Aristotle's *On Coming-to-Be and Perishing*

<2.1. THE ELEMENTS AND PRIME MATTER>

[**328b26** We have discussed how mixing, contact, action, and being affected belong to things which are naturally subject to change; also coming-to-be *simpliciter* and perishing, how they occur and to what and for what reason; similarly we have discussed alteration, what alteration is, and what differentiates <each> of them.[248]]

204,5 * * * [249] calling the coming-to-be of what is substance in the strict sense and is better 'coming-to-be *simpliciter*', e.g. of fire, and that of what is worse, e.g. earth '*something* coming-to-be'.[250] As far as 'how' is concerned, that it is by change of the substantial form. As for 'for what reason', it is because a common matter underlies them, which is subject to change and receptive of contraries. 'And what differentiates … of them.' He omits 'each', i.e. what is the differentiating feature each of them has in relation to the others.

328b31 It remains to consider the so-called 'elements' of bodies.

Either one ought to change the case and read 'It remains to give consideration to the so-called 'elements',[251] [There is also in Homer the line 'And the two rocks, the one reaches up to the broad heaven', in place of 'And of the two rocks, the one reaches up to the broad heaven'],[252] or else one must supply in thought from outside the words 'How does coming-to-be and perishing occur in them?', so that the argument may be continuous with what has been said before. For that
205,1 is how the whole argument goes if you take it up from an earlier point. We have said already, he maintains, how contact and mixing and acting and being affected belong in general to things which are subject to natural change, and we have said what it is, in general, that is

subject to coming-to-be and perishing, and it remains next that we must consider, after the general account <we have given> of them, how these things occur in the case of the so-called elements, i.e. how coming-to-be and perishing and the other things we have spoken about belong to the elements.

He says 'so-called' because it was not he himself who gave them this name but it is in circulation in common speech. For these are not properly called elements, but as in speech those things which are primary and simplest and incomposite are called elements, I mean the twenty-four letters of the alphabet,[253] so in the case of natural bodies those things which are primary and most simple would properly be called 'elements'; and these are matter and form.[254]

Or <perhaps> he says they are 'so-called' on the grounds that it has not yet been proved that these are the things which in accordance with the definition of 'matter' underlie all composite bodies subject to coming-to-be and perishing; for it is this which is his object for the present moment: to explain the manner of their coming-to-be from one another, and to demonstrate that those that have been named comprise the full and exclusive list of things out of which everything subject to coming-to-be and perishing is composed.

> **328b32** For coming-to-be and perishing in the case of all substances which are by nature composite do not occur without the perceptible bodies.

His intention is to discuss the elements and their changing into one another. Accordingly, he shows that neither they nor any one of them is what underlies coming-to-be and perishing, as some <philosophers> thought. These <philosophers'> views he <now> sets forth. For the true principles and primary elements of things subject to coming-to-be and perishing are matter and form, out of which these so-called elements are themselves composed, and coming-to be and perishing occur in them. But if they too belonged to the class of things which come to be and perish, manifestly they would not be the elements of everything subject to coming-to-be and perishing, but <the elements would be> the matter and the form which are common to these and all the things that are subject to coming-to-be and perishing.

The order of presentation appropriate to teaching is of that sort. But what he says needs stating is something like this. Since, he says, the coming-to-be and perishing of all natural substances is not without perceptible bodies,[255] but always <takes place> in these, it would be reasonable for someone who is looking for the principles of things that come to be and perish to look for the principles and elements of perceptible bodies; for <it is> amongst these <that> the things that come to be and perish <are to be found>.

But notice how as a precaution he avoids saying that all the things 206,1 that come to be and perish are perceptible bodies, which someone might unthinkingly suspect, believing that the things that come to be and perish are necessarily perceptible bodies, although perceptible bodies are not in all cases capable of coming-to-be and perishing, for instance the celestial bodies; but he considers the matter carefully, and, lest anyone should cite the natural powers, which are physical substances and come to be and perish, but are not also perceptible bodies, he puts forward a thesis which is at the same time entirely correct and thoroughly cautious, maintaining that coming-to-be and perishing do not occur in the case of substances which are naturally constituted without perceptible bodies. For the natural powers are themselves too substances, even if they are not bodies, but they have their being in bodies and are brought into existence and perish in them.[256] So they do not participate in coming-to-be and perishing without them. And if in the *Physics*[257] in the books devoted to movement Aristotle says that the natural powers and the lives that are in natural bodies do not have coming-to-be, his remarks are not incompatible with what is said here. For there he called 'coming-to-be not *simpliciter*' the passage from not being to being. The natural powers and the lives <of things> do not have this sort of coming-to-be, but in so far as they are brought from not being to being and again from being to not being in any way at all, in this sense they are said to come to be and perish.

328b34 [The underlying matter of these] is said by some to be one, and they posit air, for instance, or fire, or something midway between these two, which is at once a body and separable.

Having argued that those who give an account of things that come to be and perish need to investigate the principles of perceptible bodies (for it is in these that coming-to-be and perishing occurs), he now proceeds to the investigation, and shows that the four so-called elements are not the common matter of perceptible bodies, nor, strictly speaking, their principle, neither all of them nor <just> one of them, as some have thought, but that they themselves are perceptible bodies capable of coming-to-be and perishing. He cites the opinions of those who believed them to be principles of things capable of coming-to-be and perishing. For some supposed that some one of them was the principle of bodies, e.g. fire according to Heraclitus of Ephesus and Hippasus, air according to Diogenes of Apollonia and Anaximenes, water according to Thales and Hippon; but Anaximander supposed that some other thing over and above the four elements was the principle of things that are.

'Or something midway between these two,' which is at once a body and separable. Since Anaximander, rightly supposing matter to be the principle of bodies, could have said that it was midway between 207,1
the elements as something underlying them (for this is how we speak of the middle term in the third figure <of the syllogism> as being the same as neither of the extremes, but as underlying them both[258]), Aristotle shows that this was not the sense in which he put forward his theory, but something between earth and fire or between air and water, and he adds 'which is at once a body and separable'. And the next thing he does is to show that it cannot be supposed that there is some simple body which subsists over and above the four. For if there is a subsistent body, it will be in space and endowed with qualities, and being in space it will be either above or below, and if above it will turn out to be either fire or air, and if below either water or earth. And the same argument holds in the case of qualities. For it will have to have one or other of the contrarieties; for it will be either light or heavy, either hot or cold, and either wet or dry. But whichever of the aforesaid differentiating features you use to characterise it, it will be bound to be one or another of the four. So having called it 'body', he does well to add 'separable', to make clear that natural body is what is meant; because mathematical <body> is not like this.

> **329a1** Others hold that the number is more than one, some naming fire and earth, [others adding air to these to make three, others again adding a fourth to these, namely water, as Empedocles did.]

Having discussed those who suppose that some one of the four is the principle of bodies, he now sets out <the views of> those who suppose more than one principle, either some of the four <or all: e.g.>[259] fire and earth were suggested by Parmenides, these together with air by Ion of Chios, the tragedian, and the four by Empedocles.

> **329a3** From the aggregation and segregation or alteration of these, they maintain, the coming-to-be and perishing of things results.

<Aristotle> has said in the previous book that, whereas those who postulate more than one principle maintain that coming-to-be and perishing occur by means of aggregation and segregation and are able to preserve alteration as something different from coming-to-be and perishing, those who postulate a single principle, since the principle has to remain unchangeable if it is to be a true principle, are forced

to speak of coming-to-be and perishing in terms of alteration and not
208,1 in terms of change of substance. And thus they are compelled to reduce coming-to-be and alteration to the same thing, being unable to come up with a difference between them.[260]

> **329a5** Let it be conceded that 'principles' and 'elements' are good names for the primary things [whose change, whether by aggregation and segregation or another change, entails that there is coming-to-be and perishing. But those, on the other hand, who postulate a single matter over and above those mentioned, and that corporeal and separable, are mistaken.]

It is reasonable, he says, to call the primary things out of which the rest come to be 'elements' and 'principles'. But to posit as primary those things which these people mention is not reasonable. And he lets this pass as something <already> implied and obvious. And some posited the four elements, either some of them or all, and others a further thing over and above these, for instance, Anaximander. And <Aristotle> refutes him, not as positing something other than the four elements as matter (for in this sense <Anaximander> could have spoken correctly, if he had said that it was incorporeal and formless), but <Aristotle> refutes him on the grounds that he supposes it to be a separable body.

> **329a11** [For it is not possible for this body to exist without a perceptible contrariety] – for this infinite must necessarily be either light or heavy, or cold or hot.

Having shown that if the matter which Anaximander posits is a subsistent body it will be necessary to consider it as having a contrariety, although the additional premiss is needed 'But if has a contrariety it will be one of the four', he leaves this to be supplied in thought as being obvious.

> **329a13** The account given in the Timaeus, on the other hand, lacks precision. [It does not say clearly whether the omnirecipient is separated from the elements …]

From this he turns his attention to Plato, and first he accuses him of saying that matter was the principle of natural things without clearly determining whether it could be separated from the bodily elements as subsisting by itself or not. And it makes no difference whether by 'elements' he means the four or the cubes and pyramids which Plato posited; for the argument against him is <to demand> whether matter can subsist in its own right or whether it is associated with

some body. And he says that Plato says nothing clear about this, but it should be realised that even if Plato did not speak so clearly nor using the same terminology as Aristotle on this topic, nevertheless what he says is sufficient to make clear his own opinion on this. For it is he himself who says in his discussion of matter, 'For it is always receiving everything, and the form allotted to it is not like any of the 209,1
things that come into it.' But if it is always receiving, it is evident that it never subsists in its own right.

> **329a15** ... and he makes no use of it, having said that it is a sort of substratum to the so-called elements [as gold is to artefacts made of gold. (What is more, this is not well said, said in this way: things are like this in the case of alteration, but in the case of coming-to-be and perishing it is not possible to call something by the name of the thing from which it has come to be. He says, however, that it is 'far and away the most true thing to say' that each of them is gold) ...]

Now he launches another accusation against Plato, namely, that after postulating that there is a matter which is a substratum for the elements he makes no use of it in his cosmology in connection with the coming-to-be of bodies, saying that everything comes to be from surfaces being put together. And he substantiates both these claims, (a) that he posits matter as a substratum for the elements, reminding us of the model that Plato sets up[261] in the phrase 'as gold is to gold artefacts'; then, before he brings forward evidence to support the second claim, namely, (b) that he does not use matter in <his account of> the coming-to-be of bodies, he criticises Plato for another fault: because Plato says that we ought rather to apply to the things subject to coming-to-be and perishing names drawn from the substratum and matter, but it is right to characterise things that are, not from their matter, but from their form; for it is this which marks out the peculiar nature of each thing.

In reply to this, then, it might be said by someone putting forward an argument on Plato's behalf that he himself directs[262] that each thing should be characterised in accordance with its form. For he says, according to Aristotle, that each thing possesses being by means of its form, coming-to-be by participating in its form, and perishing by losing its form. But he says that it must get its name from the matter, not because it is in accordance with this that things that are enmattered and composite are characterised, but on the grounds that it is safer to call them after their matter; for when the form changes this always remains. It is after all no longer true to call the gold statue a statue if it changes and becomes a tripod, but it is true to call it gold even after the change; for the matter remains the same while *receiving* the forms.

In reply to this Aristotle might justly say that in the case of alteration one does well to name the thing that changes from its substratum, because the substratum, being already possessed of form received its character from its form and not from the affection that accrues to it in virtue of the alteration, but in the case of coming-to-be and perishing that which has come to be cannot be said to be that which it was before it came to be, and that which has perished cannot get its name from the matter which remains and has not perished. For example, when composite things come to be and perish, if someone were to ask about one of them, supposing it to happen that a horse perishes and a man comes to be, 'What is it that has perished and what is it that has come to be?', if I were to say that it was matter, I should be saying what was directly opposite to the truth; for it is not the matter that has perished (for it still remains), nor is it possible to call that which has come to be 'matter'; for that was there even before the coming-to-be. But in the case of alteration we call both that which loses the darkness and that which takes on the paleness 'a man', not 'a pale' and 'a dark'; for 'pale' and 'dark' are indicative of affections and not of a substance.

329a21 ... but he takes the analysis of the elements, which are solid, as far as surfaces, [though it is impossible for these surfaces to be the 'nurse' and the primary matter.]

Having broken off in the middle <of his argument> to deal with the suggestion that names should be imposed on the basis of the substratum, <Aristotle> now blames <Plato> for maintaining that the elements are solid and analysing them as far as surfaces, and <Aristotle> uses this as support for what he said previously, namely, that <Plato> makes no use of matter; for he proceeds, in Aristotle's view, as far as surfaces, from which he claims bodies come to be, but these surfaces could not be matter. But it must be recognised that Plato was not so weak at geometry as to think that body could be dissolved into surfaces: rather he was talking of physical surfaces, i.e. corporeal ones, which had depth as well. And clearly things like this also have matter observed in them; so in Plato's view the things that come to be are not without matter.

329a24 Our view is that there is a matter of perceptible bodies, [but that this is not separable but is always together with a contrariety, from which the 'elements' come to be.]

Now <Aristotle> sets out his own view, according to which he will appear indeed to differ from Plato, saying that we posit some sort of matter underlying perceptible bodies, never subsisting in its own right, but always observed with some contrariety; and this matter, he says, receiving into itself the contrarieties that occur in the definition of form[263] gives rise to the elements.

> **329a27** A more precise account of them has been given elsewhere.

He treated of matter and contrariety in more detail in the first book of the *Physics*,[264] where there was also a discussion by him of privation.

> **329a27** Nevertheless, since this is the way in which the primary bodies are from the matter, we must give an account of these also … 211,1

The sentence can be made clear by paraphrasing it in this way: 'Nevertheless, even if this is the way in which the primary bodies arise from the matter, we must give an account of these also', so that what he is saying is that, even if it has been shown that the elements arise from matter in virtue of its receiving contrarieties, this is not sufficient for purposes of instruction, but a precise account must be given about these contrarieties too – an account of what sort of things they are and how many – the addition of which to matter completes the four elements, which he calls 'primary bodies'.

> **329a29** … regarding matter, certainly, as a principle that is really first [which, though inseparable, does underlie the contraries (for neither is the hot matter for the cold nor the latter for the hot, but the substratum is matter for them both); so first that which is perceptible body in potentiality is principle, and secondly the contrarieties (I mean, for example, heat and cold), and only thirdly fire and water and the like.]

Having said that an account must be given of the contrarieties, he decides first what the principles are. The principles he is looking for now are principles as elements, not principles of action, and he says that matter is the first principle and contrariety, which is the enmattered form, the second. He says that matter is the first principle because, whilst it is imperishable, the contrariety is perishable (for it is in respect of this that perishing actually takes place), and that matter underlies both the contrarieties. For one of the contraries cannot become substratum for the other, e.g. the hot for the cold; for matter is the substratum of both. For as he has said in the first book of the *Physics*,[265] none of the contraries becomes receptive of its own contrary while remaining and being preserved; for the contraries are

destroyed by each other. So naturally matter is the first principle as substratum for the enmattered form (for it is not <first> in terms of value, for the form has greater value), whereas the form, which is the contrariety, is the second, as existing in the matter (for matter, which he also calls 'perceptible body in potentiality', is never separated, he says, from the forms; for being according to its own rationale actually incorporeal it is potentially body, not mathematical but perceptible, i.e. physical, body), and the third principle is the simple body, e.g. fire and earth and each of the other primary bodies, which are not a principle *simpliciter*, but in relation to composite bodies. This is why it is reasonable to posit them as a third principle, both because they are composite[266] and because they are changeable, whereas that which is the principle in the strict sense and *simpliciter* is obliged, in virtue of the fact that it is not derived from other principles, to be also unchangeable;
212,1 for that which is changeable has to be composite as well, given that the change happens to some part of it[267] which persists.

329b1 [For these change into one another,] and it is not as Empedocles and others say, for there would be no alteration.

If the elements which are substantiated[268] in accordance with the affections are unchangeable in nature, as Empedocles believes, it is clear that the affections are unchangeable and inseparable; but if so, there is no alteration, given that alteration is change of the substratum, which is endowed with form and which persists, in respect of affections. And it is not only alteration that Empedocles does away with, saying that the elements are unchangeable and bestowing on them the substantial[269] qualities, but also coming-to-be in the strict sense, change in respect of form; for this too is a change of matter in respect of the affections.

329b2 But the contrarieties do not change.

He shows that it is reasonable to take the simple bodies to be a third principle and the contrarieties a second by appealing to change, since the simple bodies are changeable (for they have matter which persists and receives the contraries in turn), whereas the contrarieties do not change. For coldness does not persist and receive heat; for contraries could not be at the same place at the same time. So it is not the contrarieties which change, but the bodies which have the contrarieties change in respect of them.

329b3 But none the less even so we must discuss what sort and how many of them are principles of body. [The others posit them and make use of them but have nothing to say about why these are they, or this many.]

We have maintained, he says, that the contrarieties are principles, and that they do not change, but even so we must state the number and the character of the contrarieties which it is necessary to postulate as principles of bodies. For it is not sufficient to say that the contrarieties are principles, but we must also enquire of what sort they are and how many there are of them. For the others who have said that some of the contrarieties are principles have written no demonstration of the necessity to posit these contrarieties and this number of them.

<2.2. THE PRIMARY CONTRARIETIES>

329b7 Since what we are trying to discover are principles of perceptible body, that is, tangible body, [and the tangible is that of which touch is the sense, it is obvious that not all the contrarieties make forms and principles, but only those that belong to touch.] 213,1

Having said that the principles are three, first matter, second the enmattered form which is the same as the contrarieties, and third the simple bodies which he says come to be out of matter when it takes on the contrarieties, he now proposes to enquire what are the specifying[270] contrarieties which matter takes on to produce the simple bodies.[271] But since postulating these particular contrarieties and not some others should not be a matter of mere assertion without proof – a thing he blames others for having done – he begins their investigation very systematically and by demonstrative reasoning, virtually saying this: Since we are looking for the principles of body that is not eternal but capable of coming-to-be and perishing, and this has qualities (for it is not possible for body without qualities to be capable of coming-to-be and perishing; for coming-to-be and perishing is change of matter in respect of the affections; moreover every physical body also has movement, and movements take place in accordance with weight, I mean heaviness and lightness, and these are qualities), we are therefore looking for principles of body endowed with qualities. But this is perceptible <body>; for the senses are receptive of[272] the qualities and each thing that has qualities is in these respects subject to perception.[273] So since it has been established that we seek the principles of body that is physical and perceptible, it is necessary to investigate perceptible contrarieties, and not just any perceptible <contrarieties> but those which are nearest to matter[274] and most common and spread through all bodies.[275]

And we shall discover which these are by recurring to the senses. For it is evident that we should suppose those perceptible differentiating features which are subject to the commonest of the senses, as being the commonest, to be the specifying principles of the <simple> bodies. Of the senses, then, sight is first as the most valuable in esteem, but not as having permeated[276] through all things capable of perception. It is the same with hearing and with all the others except touch; for this is the commonest of all, and it is reasonable to call it first in this sense, in virtue of the fact that its destruction carries with it that of all the others, but not vice versa.[277] For the things which participate in any of the other senses participate also much earlier in touch, but those <that participate> in touch do not necessarily <participate> also in any of the other senses, as the zoöphytes which have touch but none of the other senses make clear.

And we shall find that the same applies to the perceptible differentiating features. The visible ones, e.g. colours, do not spread through all the bodies; for air is not coloured, or indeed visible at all. Similarly the audible ones are not either, e.g. sounds; for not everything is able to give out sound (for wool is not, nor sponges), but <only> solid things, and these not in themselves, but in the collision between them and the air. Thus if you move water while submerged you produce no sound. Nor for that matter smells; for air is odourless as well as colourless, and this is how it comes to be a transporter[278] of colours and smell. Thus the qualities of taste also are not in everything: water, for instance, is tasteless while serving as substratum for tastes. But the only differentiating features we see as having permeated through all bodies are the tangible ones; for no body is without its share of heat or cold and of dryness or wetness. So the tangible contrarieties are the commonest and have permeated through all bodies and their destruction carries with it that of all the others, but not vice versa.[279]

So these must be accepted as principles. But since they are many and different, it is necessary to set out all of them, and of these to take the more universal which are not included in <any of> the others, and not all the more universal, but <only> those that are capable of acting and being affected, since what we are looking for are the principles of bodies capable of acting upon and being affected by each other and in this way producing change in one another. He sets out, therefore, some seven contrarieties, hot and cold, dry and wet, heavy and light, hard and soft, viscous and brittle, rough and smooth, coarse and fine.

And we ask why he omits mention of the opposition between rare and dense. The interpreter Alexander says that he reduces this to heavy and light; for the rare is light, just as the dense is heavy. But it is possible to reduce them to the same thing as the fine and coarse; for fineness as a quality, if it is not thought of in terms of magnitude seems to be the same as rareness. For we call a body that is easy to penetrate fine, and this seems to be the same as rare. Thus the coarse too <is the same as> the dense; for the coarse, consisting of large particles, will not pass through all and any sized passages, but <only> through large ones. And this in a way is what the dense is like, resulting from the compression of a large amount of body into a small bulk.

Having set out all the oppositions, he weeds out those which imply nothing about acting and being affected as well as with those which are included in the others. On the one hand, heavy and light and rough and smooth <are weeded out> as not implying acting or being affected;[280] for light and heavy are not spoken of in the same way as heat and cold are in virtue of their heating and cooling neighbouring objects; for <heavy> does not make juxtaposed objects heavy. And similarly neither does light <communicate itself to juxtaposed things>. And in the same way neither do rough and smooth. On the other hand, he throws away fine and coarse, soft and hard, viscous and brittle, as included <in other things>. This is reasonable; for if we take on board the more universal contrarieties and do not throw away those that reduce to them, we shall have taken on board more than we need. And he reduces these three contrarieties to wet and dry, defining what each of these is by describing wet as being easily bounded by a <boundary>[281] not its own whilst being bounded with difficulty by its own <boundary>, and dry as being bounded with difficulty by a <boundary> not its own, but easily bounded by its own <boundary>. If therefore the wet is easily bounded by a boundary not its own and is able to fill things in that it readily fills the passages, and that which is fine-grained is also of this sort (for it is able to fill things), it is reasonable that it should be reduced to the wet. And by the same argument the coarse also <is reducible> to the dry. And the viscous, he says, is wet affected in some way. And the brittle similarly with respect to dry. In this way too, he says, the soft should be reduced to the wet; for soft is that which is yielding to touch and this is a special property of the wet. Having thus weeded out these things, he accepts the two contrarieties, hot and cold and wet and dry, singling these out as specifying principles of the elements, since they are capable of acting and being affected, and moreover are the most common and neither of them is reduced to the other.

Since what we are trying to discover are principles of perceptible body, that is tangible body. Since, he says, we are trying to discover the principles of perceptible body (for body that is capable of coming-to-be and perishing is of this sort), and every perceptible body is tangible (for it is not universally visible or audible or any of the other <things of this sort>), we must look for the specifying differentiating features of bodies that are accessible to touch.

> **329b10** For it is through a contrariety that they differ, through a tangible contrariety. [This is why neither whiteness and blackness, nor sweetness and bitterness, nor, equally, any of the other perceptible contrarieties, serve to make an element. It may be said that sight is prior to touch, and that accordingly its substratum is also prior. (It is not, however, *qua* tangible that it is an affection of tangible body …]

The primary bodies, he says, the principles of which we are trying to discover, differ from each other in respect of tangible contrarieties; and this is how he shows that it is reasonable for us both to say that perceptible things are tangible and to suppose that their principles are tangible contrarieties. For the fact that tangible contrarieties permeate all bodies is evidence for the claim that the primary bodies which are elements for all other bodies differ from each other in respect of them. So we should not select differentiating features which are visible or audible or perceptible by taste or smell; for it is not these
216,1 which specify[282] the primary bodies. For, he says, even if sight is prior[283] to touch and visible things to tangible things, which he calls substrata[284] of the senses, nevertheless they contribute nothing to the specification of tangible body, the principles of which we are now trying to discover.

> **329b15** … but in virtue of something else, [even if in fact it is prior by nature.) We must first pick out from amongst the tangible qualities themselves which are the primary differentiae and contrarieties. These are the contrarieties that belong to touch: hot-cold, dry-wet, heavy-light, hard-soft, viscous-brittle, rough-smooth, coarse-fine. Of these heavy and light are not capable of acting or being affected. They are not said of things in virtue of their acting upon something else or being acted upon by something else. The elements, however, have to be capable of acting upon and being acted upon by one another, since they mix and change into one another.]

The visible quality, he says, even if it is prior to the tangible quality, is nevertheless not an affection of tangible body in so far as it is tangible but in virtue of something else, e.g. in so far as it is visible. Sight is prior, he says, to touch, <but> not *simpliciter*; for in a way touch is prior in virtue of the fact that its destruction carries with it that of sight, but not vice versa,[285] but sight, he says, is perhaps prior by nature, i.e. as being more perfect in esteem, since it is given to the perfect animals.[286]

For even if dry and wet act to some extent upon each other by holding one another together and providing a boundary, nevertheless it is when they are mixed in with each other that they act.

For this reason scientists are accustomed to calling these qualities more material and hot and cold active, on the grounds that they exhibit their own power while remaining external to the things on which they act. But it is as well to remember the general remark that was made to the effect that we see more clearly in the case of the coming-to-be of composite things that the one opposition is capable of being affected, the other of acting. For in the manufacture of pottery too wet and dry come together and produce the matter of the pottery, cold solidifies this, and heat above all changes it and makes it pottery. This then <is what happens> in the case of composite things. But in the case of their coming-to-be from one another they are all alike capable both of acting and of being affected.

329b26 For heat is that which aggregates things that are of the same kind; for the segregating which they say fire does is the aggregating of things of the same type [for this results in foreign bodies' being expelled …]

Heat, he says, which has the capacity to aggregate substances of the same kind, is thought to have the capacity to segregate, because in the course of bringing together and uniting things of the same kind it necessarily segregates and separates different kinds of things. For the very segregation of different kinds of things which it produces is achieved on account of the aggregation of substances that are alike; for in segregating the diverse kinds of substance from one another it accomplishes the union of each kind in its own right. That is why we define the hot as that which segregates different kinds of things whilst aggregating things that are alike. And it is possible, perhaps,
218,1 to say that heat in accordance with its first definition has the capacity to segregate, if we understand segregation, not as separation (for it achieves this in the case of things of different kinds and of things that are mixed, but not in the case of unmixed things), but if we conceive of segregation in a universal sense, in the sense exemplified by melting and rarefaction. But according to the second definition this involves the separation of different kinds of things and the uniting of similar things; for when melting occurs each thing proceeds to its own place and clings to that, as happens when metals are cast, for everything that is soluble joins together and sticks to itself separating from what is earthy and insoluble.

And it is clear from this that it is melting that is a cause of the segregation of things of different kinds, not fire in its own right or primarily. For on the occasions when there are two substances which admit of the same melting,[290] fire does not segregate them from one another, but on the contrary aggregates them and unites them to each other so that they are no longer able to be separated, as takes place with regard to wax and pitch; for whilst these are segregated from one another and different it unites them, and once united they cannot be separated. If, however, there should be some earth with the wax, since it is not in the nature of both these substances to be melted, but of the wax only, the wax is unified in the course of the melting and is segregated, while the earthy content sinks to the bottom because of its weight. And when there are two substances, both of them soluble, but one more so and the other less, these on coming together are unified by the fire, but not in such a way as to be no longer capable of being segregated, as happens in the case of gold and silver and similar <substances>; for these are united with each other and again segregated because they differ from each other in their density.

But if it were maintained that he uses the expression 'of the same kind' to signify things that can be dissolved and melted in the same way as each other, or those that cannot, so that 'similar' refers not just to species but to genus, e.g. so that we should not only say that gold is similar to gold, but also things which can be melted in the same way as it can, e.g. silver or bronze, even so the statement would not have universal application. For it is not the case that fire unifies insoluble things which are <in this sense> of the same kind as each other, and on the other hand it segregates things whose nature is to be dissolved, as it does gold from silver. For it applies only to those things it unites in a way which prevents them ever again being segregated, things which have the same melting[291] in the sense that one is not more meltable than another, like lead and tin or wax and pitch and things which are extremely soft.

329b29 … and cold is that which gathers and aggregates indiscriminately things that are related and things that are not of the same type.

For amongst the things solidified by cold it is possible to see that
substances of different kinds are contained; thus straws and certain 219,1
small animals are frozen in ice.

329b30 Wet is that which is not bounded by any boundary of its own but is easily bounded, dry is that which is easily bounded by a boundary of its own, but is hard <for other things> to bound.

Having shown by means of the definitions of 'hot' and 'cold' that there is good reason to call them capable of acting, he next avails himself of the definitions of 'wet' and 'dry' to show that these are capable of being affected; for 'easily bounded' and 'hard to bound' are names of affections.

A doubt arises in connection with the definition of 'dry' about what he means by 'easily bounded by its own boundary'. Look at ash, they say; even though it is dry, it flows away. Against this objection it has to be said that even if it is not the case that ash is bounded and remains preserving its own boundary, nevertheless it can be piled up into a heap, which is not a feature of that which is wet; and moreover it has a certain consistency of its own, and does not completely slip away, as is true also of sand. Some have raised a doubt too about flame, on the grounds that, despite being dry, in ovens its configuration is due to the boundary of something else. But it has to be said first that the configuration of a flame is never determined in detail by the things which contain it, since it is upwardly mobile and does not receive its configuration from the whole of that which contains it; for if that were the case it would too quickly be extinguished, not being held up on both sides by void spaces – void, I mean, of fire. Moreover, even if it *were* to agree to be bounded by the things which enclose it, it would nevertheless be contained by its own shape; for it does not slip completely away and run off, but remaining continuous moves in an upward direction. That which is wet, however, receives its exact configuration from the containing objects; so if water is frozen in a jar and you then break the jar, you will see that it has received its configuration from the jar.

329b32 Fine and coarse, viscous and brittle, hard and soft, and the other differentiating features are from these.

Having shown that of the two oppositions the one is capable of acting (active) and the other of being affected (passive), he means to reduce all the remaining ones to these <two>, so that the two of them, the hot/cold one and the wet/dry one, are the primary ones, in order that having demonstrated this he can go on from here to show that the four elements come to be. He adds 'and the other differentiating features' either because <he is thinking> of smoothness and roughness, which before he had included in his tally, but now leaves out, or rather because of the ones he adds; for his next move is to make distinctions in respect of dry and wet and show that each of these is used in more than one sense.

329b34 Since ability to fill things belongs to wetness, because it has no boundaries but is easily bounded and follows the thing with which it is in contact, [and the fine is able to fill things (because its particles are fine and that which has small particles is able to fill things; for the whole of it is in contact with the whole <of the other>, and that which is fine is very much of this sort), obviously fineness belongs to wetness and coarseness to dryness.]

He shows how fine is reduced to wet. For fine, he says, is able to fill things; for being fine-grained it fills up the empty places. For if you put fine flour into a medimnus[292] of nuts it will penetrate into the intervening spaces between the nuts which are void of solid body and fill them up. That which is wet does the same, e.g. water; for being easily bounded it will follow, he says, the external boundary of the thing which is touching it and containing it, and it will touch it at every point. And that which is fine is also like this. So the fine is said with good reason to belong to the wet. Its contrary, therefore, the coarse, will fall under the contrary <of the wet, namely> the dry; for contraries belong to contraries. For in the same way as the dry is resistant and does not follow the things that contain it or touch their entire surface (for it is not easily bounded by a boundary not its own), so too the coarse, being coarse-grained, does not penetrate into every space.

330a4 Again, viscousness belongs to wetness; for viscousness is wetness affected in some way, e.g. oil; [and brittleness belongs to hardness, because being brittle is being completely hard – so hard as to have solidified from lack of wetness.]

He gives a succinct proof that viscousness falls under wetness, saying that what is viscous is a particular kind of wet affected in some way. For those wet things are viscous which are not easily divisible and hard to detach from the things they are in contact with and slippery, like gum, oil, and liquid pitch. And brittleness falls under dryness; for that which is solidified through lack of moisture and is on this account easy to break is called brittle.

330a8 Furthermore, softness belongs to wetness; for what retreats into itself and does not move away, which is what the wet does, is soft – [that is why the wet is not soft but softness belongs to wetness; and hardness belongs to dryness, for that which is solid is hard and that which is solid is dry.]

Again, he reduces softness to wetness and hardness to dryness; for that which retreats into itself in the face of things which come in contact with it, which is what the wet does, is soft; for this too retreats. But what differentiates it? Here again it is that softness is wetness with a certain affection, namely, not moving away, an affection not

221,1 belonging to water (for it does move away), just as viscousness was wetness affected in some way. So softness is like this too. Thus, if something is soft it is in a way wet (for it is easily bounded by a boundary not its own), but it is not the case that if something is wet it is soft. So wetness extends further <than softness>. That is why softness is also reduced to this; for what differentiates <species of> wetness is moving away and not moving away, but it is a peculiar property <of wetness> and something which is essential to it that it is not easily bounded by its own boundary but easily bounded by a boundary not its own. So, again, hardness too is reduced to dryness. For what is hard is solidified (for it is made hard by its parts being compacted and solidified), and that which is solidified is hard; for by being solidified it is made hard to bound by a boundary not it own, so it is made dry.

> **330a12** 'Dry' and 'wet' have more than one sense, Both 'wet' and 'moist' are opposed to 'dry', and conversely both 'dry' and 'solid' are opposed to 'wet'. [All these belong to the dryness and wetness that are primarily so called. Since 'dry' is opposed to 'moist', and that which has alien wetness on its surface is moist (whereas that which has it in its depths is sodden, and that which is deprived of this is dry), obviously moistness will belong to wetness, and the dryness which is opposed to it to dryness in the primary sense. Again, 'wet' and 'solid' behave in a similar way: 'wet' <i.e. liquid> is that which has its own wetness in its depths (whereas that which has alien wetness is sodden), and that which is deprived of this is solid: accordingly these too belong, the one to dryness, the other to wetness.]

Once Aristotle has shown that all the tangible oppositions are reducible to the two primary ones, the hot/cold and the dry/wet, he wishes to prove that certain other <oppositions> besides those he has already enumerated are reducible to the dry/wet, <reducing> to the primary sort of wetness moistness and wetness in the more specific sense and to dryness solidity and another more specific sort of dryness which shares the same name as what is dry in the more generic sense. And in order to show this he says that 'wet' and 'dry' have more than one sense. And in showing that these <words> have more than one sense he uses the argument from opposites; for when the same thing has more than one thing opposed to it, it too is something that has more than one sense. For example, opposed to the heavy there is in sound the sharp[293] (for the sharp sound is opposed to the heavy sound), and in weight the light; for again the heavy and the light are amongst the opposites in things that have mass.[294] So since both the light and the sharp are opposed to the heavy, 'heavy' is a word that has more than one sense. But 'sharp' too is a word that has more than one sense; for 'blunt' and 'sweet' and 'heavy' are all opposed to it, 'heavy' in sound, 'sweet' in taste, 'blunt' in shapes or simply in things that have mass; for we talk of a sharp and a blunt angle[295] and of a sharp and a blunt knife. So in this way he shows that both 'wet' and 'dry' are words which have more than one sense; for the opposites of each of them are more than one. For we sometimes oppose 'wet', sometimes 'moist', to 'dry'; and to 'wet' <we sometimes oppose> 'dry' and sometimes 'solid'.[296]

There is, then, the primary and, as it were, most generic sense of 'wet', the definition of which he has already given, namely, 'that which is not easily bounded with its own boundary but easily bounded by a boundary not its own', and the opposite of this is 'dry', I mean 'dry' in its primary and most generic sense, namely, 'that which is easily bounded by its own boundary but not easily by a boundary not its own', as he earlier defined it. And he reduces 'moist' and 'wet' in the specific sense to 'wet' in the primary sense: 'moist', then, is <the name for> that which has alien wetness on its surface, e.g. if you pour a modest amount of water on to a stone or our body or clothing (for the name for that which has alien wetness in its depths is 'sodden'), whereas 'wet' is <the name for> that which has its own wetness in its depths. So moistness differs from the primary sort of wetness in that that which is wet in the primary sense is wet through and through in virtue of the very fact of being such, but that which is moist <is wet> in virtue of having some affection of this sort; and that which is wet *simpliciter* differs from that which is wet in the specific sense in that that which is wet in the primary sense is such through and through, as I have said, whereas the other has its own wetness hidden in its depths. But 'wet' in the specific sense differs from 'sodden' in that the one has its own, the other an alien, wetness in its depths. An example of something sodden is mud, but examples of what is wet are things that can be melted, wax or lead or that sort of thing.

Again, he reduces to dry in the primary sense both that which is dry in the specific sense, which is opposed to what is moist and which shares the same name with the genus (it happens not to have a name of its own), and that which is solid, which is opposed to what is wet in the specific sense, which itself too shares the same name with the genus because it happens not to have a name of its own. So since moist is what has alien wetness on its surface, the dry opposed to this will clearly be that which lacks the sort of wetness which the moist was said to have. The same applies also to 'wet' in the more specific sense and 'solid'; for the former has its own wetness in its depths, whereas the latter lacks this, as, for example, things that cannot be melted like shells and those minerals which are unmeltable; and meltable things too, before they have been melted, are solid. Solid things therefore are divided into those which can be melted and those which cannot, and those which can are wet when they have been melted and the wetness that is in their depths comes into the open. But if these things get their names from the absence or presence of wetness, it is reasonable that they should be reduced, the one class to what is wet in the primary sense, the other to what is dry in the primary sense.

He brings in 'sodden', according to Alexander, in order to clarify 'moist' and 'wet' in terms of what differentiates them from it, to lend credence to the view that each of them is such as he has defined it; for as I said, what is sodden differs from what is moist in having alien
223,1 wetness in its depths whilst the latter has it on its surface, and from what is wet in the specific sense by having alien <wetness>, whilst the latter has its own wetness in its depths.

> **330a24** Manifestly, then, all the other differentiae are reducible to these four primary ones, whereas these cannot further be reduced to a smaller number. For neither is heat the same as wetness or as dryness, nor is wetness the same as heat or as cold, nor are cold and dryness fall either under one another or under heat and wetness. [So of necessity there are these four.]

We have, then, shown by means of the explanations already given how the other oppositions are reduced to the two. Alexander, however, raises a difficulty: how is smoothness reduced to wetness and roughness to dryness? It is because roughness would appear to belong to solidity and dryness (for you can see the boundaries at which its parts are connected) and smoothness to wetness (for what is wet is something uniform, with its surface always smooth, because its parts settle down together with each other). Or better, he says, to say that these contrarieties are neither active not passive, since in origin they are not even qualities. For rough and smooth, as he says in the *Categories*,[297] consist in a certain arrangement of the parts. But if they are not qualities, they are neither passive not active.

And he explains that neither of the two most general contrarieties, the hot/cold and the dry/wet, is reduced to the other; for neither does dry fall under hot (since everything dry would have to be hot; but, as it is, earth, as well as being dry, is cold), neither does wet, for that matter, fall under hot; for water, as well as being wet, is cold. Much less, then, is it the case that cold falls under hot; for how could antagonists fall under one another? Again, neither does dry fall under cold (for fire, as well as being dry, is hot), nor does wet (for air is wet and hot), and much less its contrary, hot. And for the same reasons, even if we reverse the order, we shall not reduce the others either to wet or to dry.

<2.3. THE CONTRARIETIES AND THE ELEMENTS>

330a30 Since the elements are four in number, and of the four the pairings <are six>, [but it is not in the nature of contraries to be paired with one another (it is impossible for one and the same thing to be both hot and cold, or, again, wet and dry), obviously the pairings of the elements will be four in number: hot and dry, and wet and hot; and, again, cold and dry, and cold and wet. And they are attached correspondingly to the apparently simple bodies, fire, air, water and earth. For fire is hot and dry, air hot and wet (for air is something like steam), water cold and wet, and earth cold and dry. So it is in a rational way that the differentiae are allotted to the primary bodies, and the number of them corresponds.] 224,1

He calls the contrarieties elements (for the contrarieties were said to be elements and principles in second place after matter), and he says that of the four contrarieties six pairings are produced, and of these those which take contraries are not combinable, the hot and the cold and the dry and the wet (for it is impossible that contraries should exist together), whilst the four others can be combined and it is, he says, only to be expected that the apparently simple bodies should correspond to them in number; for these also are subject to perception each one of them being characterised by a particular pairing, fire by hot and dry, air by hot and wet, water by cold and wet and earth by cold and dry. And it should be noted that he posits the others without a proof, but in the case of air alone, as giving rise to some uncertainty, he makes use of a proof of its being hot and wet, not proving that it is wet, but that it is hot; for there have been those who held the opinion that it was cold and wet.

What then does he say? <He says> that air is a sort of vapour, and vapour both *is* hot and owes its coming-to-be to heat. And there are several ways of proving this: first, from the fact that it is light, which is <a feature> of heat. And you can recognise that light is <a feature> of heat from this: for since fire, which is hot and dry, is light, it must be the case that it is light either in virtue of being dry or in virtue of being hot. But if fire were light because of its dryness, it would have to be the case that earth, being dry, was light; but this is not so; therefore it is not because of its dryness that it has lightness. It remains therefore that it is because of its heat. So if lightness comes to be by reason of heat, and air is light, it will also be hot itself. Moreover, we see that the elements that are each other's neighbours have a certain amount in common with one another, since <other-

wise> they would not even be near each other. So if air differs from fire in respect of wetness, it will evidently be associated with it in respect of heat; and if it is associated with water in respect of wetness, unless it differs from it in respect of heat there will be no difference between water and air. Moreover, we see that change into it from water takes place by means of heat; for water when vaporised is changed to air, as Aristotle himself also says.

225,1 **330b7** For of those who make the simple bodies elements, [some make them one, some two, some three, some four. Those, then, who say there is only one, and then generate the others by condensation and rarefaction ...]

After having identified the sort of contrarieties which are needed to be principles and specifiers[298] of bodies, he now intends to obtain confirmation of his own opinion on this matter, and by doing so show that all those of his predecessors who have an opinion about elements introduced the contrarieties that have been mentioned, either all or some, and never exceeded the number four; for even if they did not talk about them in the same way, nevertheless, as though compelled by the truth itself, they manifestly speak in favour of these. For not only do those who make one of the four <the real> element[299] of them all bring the others into existence out of it by rarefaction and condensation (and rarefaction and condensation are produced by heat and cold), but Parmenides also having posited originally two elements, fire and earth, brings in the contrariety, hot/cold, saying that air and water are a mixture of the two elements, as Plato also does. For he too posits a certain pair of principles, great and small, and then posits a third which he says is a mixture of the contrariety between great and small, saying that great and small are the specifying differentiating feature which supervene upon matter, which it first receives when it becomes endowed with quantity and extension, whilst he calls matter a mixture, not in virtue of the fact that it comes to be as a result of the great and the small having been mixed, but in virtue of the fact that when these are mixed in it one or other of them in turn comes to be in it. And Empedocles himself also supposes the elements to be four and introduces hot and cold as a contrariety, maintaining that fire is hot, whilst saying that all the others are opposed to this as cold.

330b11 ... for them the consequence is to make the principles two, the rare and the dense, or the hot and the cold. [For these are what operate, and the one underlies them as if it were matter, while those who make them two from the start, as

Parmenides does fire and earth, make the intermediates blends of these, i.e. air and water.]

Those, he says, who posit the one as element and say that the other things come to be from this by rarefaction and condensation make two principles of the rare and the dense, or rather of the hot and the cold; for these are the things that are operative and productive of rarefaction and condensation.[300] And it should be noted that he reduces rare and dense to hot and cold, which is perhaps why he passes over rare and dense in his enumeration of the tangible qualities on the grounds that they are reduced to hot and cold.[301]

330b15 And likewise those who mention three, like Plato in the *Divisions*, [for he makes the intermediate a blend.]

And Plato, he says, positing three principles in the same way as Parmenides, sees two in terms of the great/small contrariety and that which is between these, which he also says is a mixture of both, as third. But it must be realised that he mentions Plato in the present context, not as introducing things which serve as principles for the four that have already been demonstrated, I mean hot and cold and dry and wet, but as himself too seeing a certain contrariety among the principles, as Aristotle himself does, and moreover on account also of the similarity between his theories and those of Parmenides; for they both posit a single contrariety, and as Parmenides posited air and water as a third thing <made> out of fire and earth, so Plato too calling matter a third principle said that it was between the contraries, since it underlies both of them in turn, and for this very reason is also a mixture of both in so far as it is receptive of both.

'... like Plato in the *Divisions*.' We must enquire which are the works he calls 'the *Divisions*', where he says Plato disposes of these matters. The commentator Alexander says that the work that is in circulation as 'Plato's *Divisions*' is spurious, but, he says, Aristotle could be talking about what is said in the dialogue, the *Sophist*, calling the material in that work 'Divisions'. But it should be recognised, for a start, that there are no 'Divisions' in circulation at all whose title describes them as being by Plato, and that in the *Sophist* Plato cannot be seen to be saying anything of the sort he is saying here, nor can one find him expounding any opinion of his which is like this; for in general it is not one of his aims there to enter into a discussion of such topics, except in so far as by way of history he rehearses various views of men of science. So it is preferable to accept the final additional suggestion that Alexander makes, saying that Aristotle is speaking of Plato's unwritten opinions, which Aristotle himself had copied, and calls these 'the *Divisions*'.[302] In these, therefore, Plato posits the great and the small and that which is between them.

227,1 **330b17** For practical purposes those who make them two and those who make them three are saying the same thing, except that the former divide the intermediate in two, whereas the latter make it only one.

It is possible that this is being said of Plato and Parmenides. For as Parmenides proposed that there be two principles and made something else a mixture out of these, so Plato said that there were three, and, while regarding two in terms of the contrariety, posited a third midway between these. The two men differed only in this, that whereas Parmenides divided the middle one into two, air and water, Plato keeps it undivided. But the dictum can be understood in another way, no longer supposing him to be making a comment on Plato, but as himself utilising arguments against that <philosopher>, and saying something of this sort: that Parmenides, supposing there to be two principles, took something midway between these, which he divided into air and water, whilst others, like Ion, supposing from the start that fire, earth and water were principles, and themselves saying that water was intermediate between the other two, preserved it undivided. But they posited air and water as intermediate between fire and earth, paying attention perhaps to weight; for it is not possible to say this in terms of the qualities, given that earth and fire are dry and these other two, by contrast, are wet.

330b19 Some, e.g. Empedocles, mention four from the start; but he in fact contracts these to two, [for he opposes all the others to fire.]

Empedocles too, he says, while supposing the four elements to be four principles, nevertheless supposed that there was a single contrariety, maintaining that fire was hot and the other three cold.

330b21 But neither fire nor air nor any of those we have mentioned is in fact simple, but mixed. [The simple bodies are like these but not the same as them.]

Those who look at his words too hurriedly will think that what he is saying is that the elements are not simple or pure but mixed and always rendered spurious by each other, as many have held.[303] This
228,1 idea is worth consideration in its own right. And the next thing to investigate is how things stand in this regard – but only after we have first set out the true sense of what he says. And he says that air and

the others turn out not to be simple in the way the other principles, primarily and secondarily so-called, are <simple>; but the former, he says, are mixed and composite, having matter and form as their components. But the forms that constitute the contrarieties are, he says, simple; for the form of fire, which is regarded as constituted by the pairing of the qualities – I mean hot and dry – this is simple, just as matter is simple, but fire is already something composed out of both.[304]

Something like this then is what is meant by <Aristotle's> remark. But the question about the elements is worth investigating in its own right, whether, that is, none of the elements is pure,[305] as some people say, but all are mixed up with each other and made spurious by each other, or whether there is nothing to stop them also subsisting in their own right and regarded as uncontaminated by involvement with each other. In response to this it must be said that if we characterise fire, as Timaeus[306] does, by what is visible and earth by what is tangible, in this way everything will be in everything and nothing will be on its own; for all the elements are tangible and almost all visible. But if we characterise each in the way Aristotle does, dry by its being easily bounded and wet by its not being easily bounded and hot by its tendency to segregate and cold by its tendency to aggregate, there will be nothing to prevent their being pure and unmixed in respect of one another; and if there is nothing to prevent it, but it can happen, then it might come about.

330b23 For example, that which is like fire is fiery, not fire; [that which is like air is aeriform; and so on in other cases.]

The differentiating features that are specific[307] to the simple bodies are, he says, like the elements, but not the same, since fire resembles the pairing[308] which serves to specify it, but this pairing is not the same thing as fire, but the likeness <is> in respect of quality, and the pairing is not fire, but is fiery and is the form of fire. And the same remarks apply to each of the others.

330b25 And fire is excess of heat in the same way as ice is of cold. [Solidifying and boiling are excesses of a kind, one of cold, the other of heat. So if ice is the solidifying of wet and cold, accordingly fire is the boiling of dry and hot: this is why nothing comes into existence either from ice or from fire.]

Here he is talking about the fire that is used in daily life, which he says is an excess of boiling, just as ice is an excess of cold. This is why no living thing can be engendered by ice or by the fire we are familiar with. The reason for this is that elemental fire and water, which both 229,1 engender living things and are used for the cohesion of composite bodies, if they degenerate into excess, produce, in the case of water, ice, and in the case of elemental fire, this fire which we use in daily life and describe as 'familiar'.[309]

330b30 The simple bodies being four in number, each of the two pairs belongs to one or other of the two <places>: [fire and air belong to that which moves towards the boundary, earth and water to that which moves towards the middle.]

The discussion is about the association between the elements and their places,[310] which is this weight-factor[311] according to which things are called heavy and light. So given that there are two places, one being up, which he says is 'towards the boundary' (using the word 'boundary' for the edge of the containing and encircling <sphere>), and the other being in the downwards direction and towards the 'middle', fire and air belong to the upper place, water and earth to the lower. For 'each of the two pairs', i.e. of the two pairs one each, belongs to each of the places, the one to the upper, the other to the lower; for he says 'each' in the neuter <plural>,[312] as if he had said 'one or other of the two pairs'.

330b33 Fire and earth are the extremes and the purest; [water and air are the intermediates and more mixed.]

He speaks of fire <and earth>[313] as 'extreme' and 'pure' in respect of the weight-factor;[314] for fire has the extreme degree of the tendency to move in the upward direction, earth in the downward, whilst water and air are in a certain intermediate state in respect of this potentiality, water being less inclined to move downwards than earth, air being surpassed in upward tendency by fire.[315]

331a1 Two of them are contrary to the other two, [respectively: water is contrary to fire, earth to air, because they are constituted by the contrary affections.]

One of the components of either of the pairs is, he says, contrary to one of those of the other, fire to water and air to earth. And showing what he means by saying that one substance is contrary to another he supports his remark by saying 'because they are constituted by contrary affections'. For the composite substance is not contrary to the composite <substance> when considered as matter; for contrari-
230,1 ness is not in the composite substance in respect of its substratum, but only in respect of the specifying differentiating features, which we actually call 'contrarieties'. So when we say that substance is

contrary to substance, we mean substance in respect of its form. And in this way we say that fire is contrary to water, in virtue of its receiving its specification from contrary differentiating features. But in the *Categories*[316] he says with good reason that substance is not contrary to substance, since he is developing his account of composite substances and not proposing there that either matter or form be spoken of as substance.

> **331a3** All the same, being four each belongs *simpliciter* to one of the affections: [earth belongs to dry rather than to cold, water to cold rather than to wet, air to wet rather than to hot, and fire to hot rather than to dry.]

There are four <features> which specify body, hot, cold, dry and wet, and each of the simple bodies is characterised *simpliciter* and in the strict sense[317] by one <of these>; for whilst each is completed by a pairing of qualities, as has been said, in a more strict sense each is, he says, characterised in terms of one quality.[318] So it is clear from <what was said> there that each is occupied by a pairing of two qualities. For *qua* liable to change into one another they must participate in contraries; but *qua* neighbours of one another it is necessary that they should have something in common. For example, fire is the neighbour of air, and in so far as it is its neighbour the two will have some quality in common, e.g. heat; but in so far as either of them changes into the other, it is necessary that they should differ in respect of another quality, wet and dry. But even if each participates in two qualities, he says, it is characterised properly most of all by one, fire by heat, air by wetness, <water by coldness>[319] and earth by dryness.

And some people have found it questionable that he should have apportioned wetness to air and not to water, water being more capable of making things wet, as can be seen from the activity of both: we see that water makes us much wetter than air does. The reply that needs to be given to these people is that it is a fault of method to construct a proof from the capacity to make things wet without first defining what wet is. So let us see what wet is in Aristotle's view. <It is,> he says, that which is not easily bounded by its own boundary. But to what does this <description> more <appropriately> belong? Is it clear that <it belongs> to air more <appropriately> or <that it belongs more appropriately> to water? For it is air rather than water that is dispersed and flows away and is harder to circumscribe. For water will actually preserve a certain boundary when it is let go from 231,1
above, but air will both flow away and easily be dispersed. If, therefore, it is that which is not easily bounded which is wet, air will be wet to a greater degree <than water>.

We have to ask, finally, how it is that water manifestly makes bodies wet more <than air does>. But we see that air itself also, when it is coarsened and is coming close to being water, develops a greater power to make things wet. And this happens because water, being more coarse-grained, has a tendency to fill the passages[320] in us, while air on account of its rare consistency[321] easily flows away. And this is what we should expect to happen; for water has a natural connection with earth and is easily linked up with it – something to which its weight contributes, whereas air does not have this linkage with earth. So this is why our bodies, being earthy things and participating more in this <than in the other elements>, are made wet more by water, the nature of which is to be mixed in <with them> and not flow away as air does.

<2.4. RECIPROCAL TRANSFORMATION OF THE ELEMENTS>

> **331a7** Since it has been settled earlier that coming-to-be for the simple bodies is from one to another …

He has shown that the elements are four, and what it is that each of them is specified by. His object now is to explain that they have their coming-to-be from each other, and whether <this is the case> for all of them or not, and the manner of their coming-to-be from each other. It has been determined earlier, he says, that the coming-to-be of the simple bodies is from each other: his remarks on this subject are to be found in Book 3 of *On the Heavens*.[322] And he says this same thing here too, on the one hand appealing to the evidence of perception, and on the other constructing a logical argument. And having demonstrated this he turns to the manner of their change from one to another, explaining both how they change into one another and which sort change more easily and which sort with greater difficulty.

> **331a8** … and since at the same time they are observed, even on the level of perception, to come to be (for <otherwise> there would be no alteration; for alteration is in respect of the affections of tangible objects) …

If, he says, coming-to-be for the simple bodies is not from one another, there would be no such thing as alteration, because coming-to-be for them is by change of their tangible qualities, in so far as it is by these that they are specified; but if these tangible affections are said to be unchangeable by those who deny the coming-to-be of the elements
232,1 from one another, they deny also alteration, given that alteration is

understood in terms of the tangible affections. But we ought to think of alteration, not, as Alexander does, of coming-to-be; for the sense of the remark[323] will not be found appropriate if it is taken as saying that if coming-to-be is not from one another there will be no coming-to-be. Alteration, then, must be understood as change, in the general sense, in respect of quality, whether the quality is accidental or substantial. But perhaps someone will say, 'What does he mean by saying that alteration in the general sense takes place in respect of the tangible qualities? For the change of white into black or of sweet into bitter does not take place in respect of the tangible qualities.' Our reply to this is that all the other alterations, even if they are not regarded as being in respect of the tangible affections, are nevertheless determined by change of the tangible qualities; for colours and tastes and all perceptible qualities are determined by the way in which hot and cold and dry and wet are mixed.

'At the same time they are observed, even on the level of perception, to come to be.' This interrupts the thought; for connection of thought is something like this: that coming-to-be for the simple bodies is from one to another; for <otherwise> there would be no alteration, but the words 'they are observed on the level of perception to come to be' are inserted in the middle. And it falls within <the competence of> perception <to ascertain> that that which to begin with is cold ends up as hot and vice versa, and that which is wet dry and vice versa.

> **331a11** [... we must now discuss the way in which they change into one another], and whether every one can come to be from every other one, [or whether this is possible for some but impossible for the others.]

We not only have to maintain, he says, that they are brought into existence from one another and to explain in what way they change into one another, but also <to make clear> that each of the four is brought into existence from each, and that it is not the case that there is one of them that is unchangeable, as Plato supposed earth to be.

> **331a12** <It is in fact clear>[324] that all are by nature able to change into each other. [For coming-to-be is to contraries and from contraries, and the elements all have contrariety with each other on account of their differentiae being contraries. For some both are contraries, e.g. fire and water (for the one is dry and hot, the other wet and cold), for others only one is, e.g. air and water (for the one is wet and hot, the other wet and cold).]

He now shows that they all change into each other, and the proof is like this. Contraries come to be from one another (for they change into one another; for it is not the case that anything at all can change into anything at all, as has been demonstrated), and the elements are related as contraries, so the elements all come to be from one another and change into one another. And he shows that the elements are contraries by setting out their differentiating features, which are contraries, and it is in respect of these that they are specified and change into one another.

233,1 **331a21** [So in general it is clear that it is natural for every one to come to be from every other;] and consequently it is not difficult to see in each case taken individually how it happens. [For, while they will all come from each other, they will also differ from each other in that with some it is faster, with others slower, and with some it is easier, with others more difficult. The change is fast in the case of those which have counterparts relative to one another, slow in the case of those which lack them, because it is easier for one thing to change than for many. For example, from fire there will be air if one of its properties changes, the former having been hot and dry whilst the latter is hot and wet, so that if the dryness is conquered by wetness there will be air.

Again, from air there will be water if the heat is conquered by cold, the former having been hot and wet, the latter cold and wet, so that if the heat changes there will be water. In the same way there will be earth from water and fire from earth. For both have counterparts relative to both: water is wet and cold, earth cold and dry, so that if the wetness is conquered there will be earth; and again, since fire is dry and hot, whereas earth is cold and dry, if the cold is destroyed there will be fire from earth. Clearly, therefore, the coming-to-be of the simple bodies will be cyclical, and this is the easiest way in which change takes place – on account of the presence of counterparts in consecutive pairs of elements. Although it is possible for water to come to be from fire, and earth from air, and again fire and air from water and earth, it is more difficult because more things have to change. If there is to be fire from water both the cold and the wetness have to be destroyed, and again if air from earth both the cold and the dryness have to be destroyed. In the same way if there is to be water and earth from fire and air both things have to change.]

Having said that they are brought into existence, not some of them and others not, but each from each, he now passes to <the consideration of> a second problem, trying to discover the manner of their change from one another. And he says that some of them have a quality in common in relation to each other, calling this a counterpart.[325] For example, fire has something in common with air in respect of heat, and air with water in respect of wetness, and generally speaking, all of those that are consecutive[326] to one another. But some have nothing in common with each other, being contrary in respect of both the pairings:[327] fire is related in this way to water and air to earth. Those, he says, which have something in common with each other change easily, because they are changing in respect of one quality only, whereas those which are contraries in both respects come to be from one another more slowly; for both the qualities have more difficulty in changing than one. So in general they all change into each other, but those <which have>[328] an affinity in respect of one of the contrarieties <do so> more easily, those which are contrary in all respects <do so> more slowly.

> **331b11** This latter sort of coming-to-be takes longer. If, however, one <affection> of each member of a pair perishes, the process is easier, but the change is not a mutual one: [from fire and water there will be earth or air, and from air and earth fire or water. When the cold of the water and the dryness of the fire have been destroyed there will be air, since there remains the heat of the one and the wetness of the other; but when the heat of the fire and the wetness of the water have been destroyed there will be earth, because the dryness of the one and the cold of the other will remain.
>
> Similarly too from air and earth there will be fire or water: when the heat of the air and the dryness of the earth are destroyed there will be water, since there remains the wetness of the one and the coldness of the other; but when the wetness of the air and the cold of the earth are destroyed there will be fire, because the heat of the one and the dryness of the other will remain, and these are precisely the things which constitute fire.]

He has spoken about the change of <bodies> that possess a counterpart in relation to each other and are close to each other, and further about that of contraries, which occurs when they change in respect of both <their qualities>. This, he says, is more difficult. Now he exhibits a third variety of change, according to which from any two of the contraries[329] a third one is produced, each of them changing in respect of one of its qualities, e.g. when from fire and water which are contraries sometimes air and sometimes earth comes to be. For when fire changes in respect of dryness and water in respect of coldness, the heat of fire remains and the wetness of water, which <together> specify air; but when the heat of fire and the wetness of water perish, what remains is the coldness and dryness which are constitutive of

earth. And the same result is obtained in the case of air and earth.
234,1 For sometimes it will be water, when the heat of air and the dryness of earth perish; for what remains is heat and dryness, which specify fire.

So this process by which two elements change and give rise to the production of one other over and above these occurs in the case of contrary elements, but never in the case of those which are consecutive;[330] for it is possible for some other third <element> to be produced neither when fire and air change, nor when air and water <do so>, nor when water and earth <do so>. And for the sake of clarity let the proof be rehearsed in detail for the case of fire and air. What I mean is that these cannot change in respect of one of their qualities and produce some one <element> over and above themselves. Necessarily either the heat of fire and the wetness of air will perish and what is left will be the dryness of fire and the heat of air, which together specify fire, and there will not be something else over and above the these;[331] or again the dryness of fire and the heat of air <will perish> and the hot and wet which is left will be the same as air; or the dryness of fire and the wetness of air <will perish> and two heats[332] are left, which specify nothing; or the heat of both fire and air will perish, and there will be left dryness and wetness, which are contraries which are not combinable. And the same argument will apply to the other pairs of consecutive[333] elements. This makes it clear that there is no way in which some one third <element> can be produced as a result of two consecutive ones' changing.

The overall sense of what is said is something of this sort, but it is possible to understand the remarks differently. For, on the one hand, Alexander gives this comparatively simple interpretation of it, which accords with what Aristotle himself is manifestly saying, an interpretation which claims that contrary elements can produce something numerically one, when each of them changes in respect of one of their qualities, keeping the other. For earth and air sometimes produce fire, sometimes water. For if the wetness of air perishes and the heat remains, and the coldness of earth <perishes> so as to leave behind its dryness, fire will be produced; and again, there will be water if the heat and the dryness perish and wetness and cold are left behind. And this is why, in Alexander's view, he says that this change takes place more easily than that which produces one pair of contraries from the other one, because, he says, in the process by which fire comes to be from water two qualities perish, heat and dryness, and another two are brought into existence, wetness and coldness, but in the case of the change which we are considering, in accordance with which a single third <element> is produced out of two, two indeed perish, but different <qualities> are not brought into existence.

This, then, is Alexander's view. But it is possible, on the other hand, to hold that Aristotle says that one is produced from two, not numerically one, but specifically. For it is not possible that this 235,1 particular quality should perish without its contrary coming to be: for example, he says, as a result of fire and water changing, air is produced, not one <quantity of air> numerically, but one <coming> from fire and another from water. For when the dryness of fire perishes, in every case wetness succeeds and air is produced; and likewise when water loses its coolness and acquires warmth, air is produced.

Maybe someone will say: If we were to say that two qualities perish and two come to be, in what way would this change be said to be easier than the change to contraries, given that in either change there is a coming-to-be of two qualities and a perishing of two to be considered? To this we could reasonably reply that when the contrary element changes into its contrary the whole change is thought of in respect of the same thing, i.e. the coming-to-be of the two qualities and the perishing of the other two, whereas in the change we are dealing with now, in respect of which one thing is produced out of two, the change of the two qualities is thought of in respect of two things, each one changing in respect of one quality, fire changing in respect of dryness and water in respect of coldness. And this is the easiest kind of change, differing in no way from the first <kind of change> he spoke of, according to which of those which are consecutive and have mutual counterparts one changes into one; and that is judged to be the easiest of all <changes>.

> **331b24** The coming-to-be of fire agrees with what we see to be the case; for flame is the best example of fire, and flame is burning smoke, while smoke comes from air and earth.

Having said that a single product results from earth and air, sometimes fire, sometimes water, he says that in the case of fire this sort of coming-to-be is manifest <for all> to see. For flame, he says, which is in fact fire, is nothing other than burning smoke, and smoke is a mixture of air and earth; for the exhalation from moisture having something earthy mixed with it constitutes smoke, and when the earth has changed in respect of its coldness and the air in respect of its wetness, flame comes to be, having the dryness of the earth and the heat of the air.

> **331b26** Where consecutive members of the cycle are concerned, it is not possible that the destruction of one elementary quality in each member [should bring about a change into any of the sorts of body; for what will be left in both members is either the

> same qualities or contrary ones, and from neither of these alternatives is there room for a body's coming-to-be: e.g. if the dryness of fire and the wetness of air were destroyed, leaving the heat of both; or if the heat of each were destroyed what was left would be the contraries, dryness and wetness. Similarly too in the other cases.

As in the case of the former passage, two interpretations are possible here: one, following Alexander, denies that elements which are next to each other can give rise to just one product; for if we suppose that
236,1 one quality perishes from each of them, the remaining ones will be found to be either contraries or the same one, and it is not possible for a third <element> to be produced either from contraries or from one and the same quality, as we have shown beyond doubt at an earlier stage. This then is Alexander's view. But according to those who say that a quality cannot perish without its contrary coming to be in its place, it will not be true to say that some third <element> cannot be produced when consecutive <elements> change; for when fire and air each change in respect of one quality, there will be <resulting elements>. But if each of them changes in respect of heat, there will be earth and water, earth from fire and water from air,[334] and similarly in the other cases.

How then on this interpretation is it possible to preserve the truth of Aristotle's thought? Will it do to say that, thinking of one quality from each being left only in terms of the definition, he maintains that if these are combined together they cannot produce a third <element>, although the nature of the facts does not bear this out? For neither is it possible for the heat of fire to perish without its being succeeded by cold, nor <are similar things possible> in any of the other cases. So what is said has to be accepted as borne out in thought only, in the sense that if we take away the heat of fire and air and consider the remaining <qualities>, the dryness of fire and the wetness of air, these will be incapable of constituting anything, given the fact that they are contraries and cannot belong to the same thing. But neither would it be possible, if we took away the dryness of fire and the wetness of air, for an element to be constituted from the heat that is in fire and air; for each of these is specified by a pair <of qualities>.

> **331b33** Wherever consecutive members are concerned the qualities possessed are in one case the same and in the other contrary.

He has taught clearly why those leaving behind one quality from each of <two> consecutive <elements> don't produce something. For these, he says, have one of their qualities in common, the other contrary, but neither from opposites that have been left behind nor from the same qualities is it possible for something else to come to be.

> **331b35** So it is immediately obvious that some changes occur from one element to another [given the destruction of one quality, but a change from two to one requires the destruction of more than one.
>
> So now we have established that all the elements come to be from any, and have explained in what way change occurs from one into another.]

Concluding his argument, he recapitulates what he has said, namely, that when <one element>[335] comes to be from one <element>, something is produced as a result of one quality perishing, as in the case of those <elements> which have mutual counterparts, and when one 237,1
comes to be from two, two qualities perish. And the same is true when a contrary comes to be from its contrary; for then too two qualities perish, as has been said.

Notes

1. These are the four fundamental types of change in the Aristotelian schema: coming-to-be and perishing (change of substance), alteration (change of quality), growth and diminution (change in quantity) and local motion (change of place). See the companion volume *Philoponus: On Aristotle On Coming-to-Be and Perishing 1.1-5*, London and Ithaca N.Y. 1999, for Philoponus' discussion of these and their differences (SB).

2. It is an Aristotelian doctrine that any change can be analysed as a change of form in a 'matter' that persists through the change. When one of the elements – earth, air, fire and water – transforms into another, what actually happens is that each element is composed of a pair of qualities, hot cold, wet or dry. During transformation, one of these two qualities changes. Fire is hot and dry, air hot and moist: when fire turns to air, the dry gives way to moist. It is controversial whether Aristotle thinks that there is a quality that persists and is the 'matter' of inter-elemental change, or whether – as Philoponus thinks – formless 'prime matter' persists through the change. See the companion volume and the Appendix to Williams' Clarendon translation, *Aristotle's De Generatione et Corruptione*, Oxford 1982 (SB).

3. Williams translates understanding *kai* here, which seems to have dropped out from the text, pointed out by Alan Lacey.

4. See Philoponus' commentary on *GC* 1.1 for a full account of Aristotle's presentation and criticism of his predecessors. Aristotle divides the Presocratics into monists, who say that one element underlies all things and never really comes to be or perishes but merely takes on different forms, and pluralists, who may or may not allow that elements transform into one another (SB).

5. This refers to the view, already established in the preceding chapters, that mere juxtaposition of smallest parts does not produce alteration.

6. The Greek term *kai*, usually 'and', could be read as epexegetical, so that Aristotle's statement would read 'about the matter, i.e. the so-called elements'. This is how Philoponus reads it (SB).

7. The question is not whether 'There are elements' is or is not eternally true, but whether instances of an element come to be.

8. The term *tropê* here seems to echo the atomist claim (repeated at 127,21) that change happens by changing the shape, arrangement and position of the atoms. For more on this see *in GC* 1.2, 26,8ff. of the companion volume (SB).

9. The figure of speech by which a writer takes the part for the whole or the whole for the part.

10. I have translated *arkhai*, normally rendered as 'principles', on this occasion 'primary kinds of things'.

11. Philoponus seems to be stressing here that 'acting and being affected' does not entail that the relevant bodies are qualitatively altered, but merely that they take part in combination and segregation under the agency of various causes. Aristotle's complaint against, say, the atomists, is that they try to explain all

change by aggregation and segregation of smallest parts, and that qualitative alteration never really takes place (SB).

12. I follow here the interpretation Aristotle gives to the atomists' technical terms, *rhusmos*, *tropê* and *diathigê*, at *Metaphysics* 1, 985b16-17.

13. Here Philoponus' text lacks the *hup' allêlôn* (by one another) of the modern text of Aristotle: see below, 127,13 (SB).

14. *Phaedo* 70D-71A.

15. *Phaedo* 103A-C.

16. There are two distinct arguments concerning opposites in the *Phaedo*: one asserts that opposites come to be 'from' one another – the dead from the living – in order to argue that the converse process must also take place, and something living come to be from what was dead. Here, the issue is about the succession of opposite states in a substrate. The other argument denies that something which has a property essentially is capable of admitting its opposite: the soul, which is essentially life-giving, cannot be dead. In the chapters translated in the previous volume, Philoponus had alluded to the argument that the process of coming-to-be from opposites cannot cease (SB).

17. In Aristotelian terminology, it is more properly said that contrary qualities come to succeed one another in the same substrate (SB).

18. The fifth element, the stuff of the heavenly spheres, is quite different for Aristotle than the four sublunary elements, most markedly in not being susceptible to interelemental transformation, indeed, to any kind of change other than circular rotation. Thus, while it drives the motion of the cosmos, it is not affected in return. It isn't clear whether Aristotle intended the remarks in this treatise to apply to the fifth element. Philoponus (and Alexander of Aphrodisias, whose commentary Philoponus cites) try at a few points to show that what Aristotle says is carefully worded so that it could be understood to apply to the heavens also (SB).

19. Aristotle's view that the fifth element is not composite, as the other elements are, seems to provide a reason why it is not susceptible to being affected by the other elements. Richard Sorabji notes a discussion whether the fifth element might have no matter in *Matter, Space and Motion*, London and Ithaca N.Y. 1988, pp. 42 n. 69, 116. (SB).

20. The figure of speech which involves the transposition of words.

21. i.e. equivocally, or with different senses on different occasions. See Aristotle, *Categories*, ch. 1.

22. The Greek is *ta de aph' henos kai pros hen*.

23. The word for 'animal' in Greek, *zôion*, is also used to mean 'picture' or 'drawing'.

24. The feminine of the adjective *mousikê*, 'musical', is used to mean 'music', as an abbreviation for *mousikê epistêmê*, knowledge of music.

25. The preceding phrases are left here as *nominativi pendentes*.

26. *Phys*. 5.3, 226b23.

27. Ch. 1.

28. The synonymous are expressions used on different occasions with the same sense.

29. He has in mind the tangent, which of course *means* 'the touching thing'.

30. *Phys*. 5.3, 226b23.

31. The word I have usually translated 'be in contact', *haptetai*, is also on occasion rendered more conveniently by 'touch'. But the word I have translated 'touch' here, *thigganonta*, is a less usual, but normally synonymous word (according to LSJ, not used in pure Attic Greek).

32. By this Philoponus means the totality of the celestial spheres, everything in the universe, in fact, except the sublunary sphere which is the place of coming-to-be and perishing.

33. David Sedley has traced the history of the Stoic category of being relatively disposed in his 'Aristotelian relatives', forthcoming, Paris 1999 (SB).

34. This colourful phrase, *phthanousi ton ouranon*, is being used to make the point that rectilinear motion, continued *ad infinitum*, will inevitably at some point reach the circumference of the universe (and therefore, in Aristotle's view, be brought to a halt at that point, there being, literally, nowhere else to go).

35. *Cael.* 4.5.

36. For a discussion of the difficulty of interpreting Aristotle's remark here, see Williams' commentary in the Clarendon edition, pp. 114-15. He may mean that 'either both or just one' of the opposite qualities is true of objects that act, since the intermediates, air and water, have both heaviness and lightness. Alternatively, if the idea were to include the celestial spheres – which are neither light nor heavy – in the acccount, the point would be that only one of the pair of bodies in contact need have heaviness or lightness. To make this interpretation congruent with the nature of the celestial body, Alexander has to argue that the remark about being changed by one another does not imply mutual effect, i.e. that the celestial body is changed by the sublunary, but rather than it can mean one of the pair only acting on the other (SB).

37. *Cat.* 1b20-1.

38. The argument is complicated here by the different meanings of the term *kinêsis*, which can apply to all four kinds of change – change of place, alteration, growth and coming-to-be – but also has a restricted use for change of place specifically. Philoponus supposes that the clearer inference is from 'heavy and light' to 'capable of changing and being changed in place' and that Aristotle infers from 'capable of changing and being changed in place' to 'capable of acting and being affected'. If acting and being affected refer only to qualitative alteration, the latter inference is false when *kinêsis* is understood narrowly as only referring to change of place (SB).

39. This could only be an incidental effect of heaviness and lightness, in Aristotle's cosmos, in contrast to the Epicurean view that heavy things thrust aside the less heavy which thus move upwards. In Aristotle's account everything moves towards its proper place according to an internal principle of natural motion, and only incidentally acts on other bodies.

40. The four sublunary elements are capable of mutual interaction: it is by virtue of the fact that their properties are contrary that they are able to change. For Philoponus' eventual rejection of the idea of an unchanging fifth element, see Introduction, n. 12 above (SB).

41. Book 5.3, 226b23.

42. sc. in the sublunary sphere.

43. The phrase 'more relevant to natural science' is intended to translate the single Greek word *phusikôteron*.

44. Book 8.6.

45. Vitelli prints *pan ara ta poioun*, where *ta* must surely be a typographical error.

46. The word translated changing here is *metaballein*, which unlike the word translated 'cause to change' in the preceding passage, namely, *kinein*, is capable of taking either a transitive or an intransitive sense. Here it is intransitive, but confusingly in the next sentence it is transitive.

47. This phrase, *ouk eis pathos agei ta kinoumena*, just means that they do not qualitatively alter things they move. For the distinction between alteration as bringing something to a new state rather than actualizing an end that is already present, see Introduction and nn. 65 and 95 below (SB).

48. *di' hên ou phainetai Aristotelês poiêtikên einai legôn ton theon tou kosmou.* See below, 152,28, where Philoponus seems to be contrasting what is explicit and implicit in Aristotle. While Syrianus and Proclus thought Aristotle ought to, but didn't, make the prime mover a creator of the universe, Ammonius and Simplicius try to show that Aristotle did think that god is the efficient cause of the entire world: Simplicius *in Phys.* 1360,24-1363,24, to appear in this series in translation by Richard McKirahan, *Simplicius On Aristotle Physics 8.* See Introduction and n. 49 below; also Sorabji, *Matter, Space and Motion*, 249-59. Simplicius reports a possible objection there that Aristotle isn't as explicit that God is an efficient cause as that he is a final cause: *ho Aristotelês oukh houtôs phanerôs poiêtikon aition ton theon eipen hôs telikon, in Phys.* 1363,12-14. Philoponus takes a different tack, recognizing that Aristotle doesn't call God an efficient cause because this would be an inappropriate description of divine activity, related as it is to the notion of acting and thus of reciprocal effect (SB).

49. The Greek I have translated as 'the substance of the universe' is *tês ousias tou pantos*, which some might prefer to render 'the being of the whole'. [Philoponus prefers the term from Plato's *Timaeus, dêmiourgos*, to describe God's role as creator. While Aristotle restricts the role of the prime mover to being the first cause of motion, Philoponus wants to leave room for the idea that God is a creator. There is some debate whether this necessarily stems from Philoponus' Christian commitments or is merely an attempt to reconcile Aristotle's view with Plato's. For more on this controversy see, e.g. Sorabji (ed.), *Philoponus and the Rejection of Aristotelian Science*, London and Ithaca N.Y. 1987; Frans de Haas, *Philoponus' New Definition of Prime Matter*; H.J. Blumenthal, *Aristotle and Neoplatonism in Late Antiquity: Interpretations of the 'De Anima'*, London 1996, p. 39 (SB).]

50. This doesn't seem to be a quotation so much as a paraphrase of the preceding argument (SB).

51. I have used 'touching' here to render *thixis*, translating *ta kata thixin haptomena* as 'those that are in contact by way of touching'. *Thixis* is cognate with *thigganein*, on which see n. 32 above.

52. Though Vitelli gives a reference to a passage in Aristotle's *Categories*, which seems unrelated to the present text, it is more probable that what is meant is Alexander's commentary on that book.

53. sc. in the sublunary sphere.

54. Does Philoponus have in mind the mental image of the thing wanted which is supposed to trigger off the action directed at obtaining the thing?

55. The most that could be said of 'everything that is relational' is that if we assert that something stands in a certain relation to another thing, we are committed to asserting that the latter stands in the converse of that relation to the former: if Mark is the parent of Luke, Luke must be the child of Mark. However, Philoponus seems here to be committing himself to the stronger thesis that if we assert that something stands in a certain relation to another thing, we are committed to asserting that the latter stands in the *same* relation to the former, i.e., that all relations are symmetrical.

56. Here Williams translates the nouns corresponding to *poiein* and *paskhein*, 'acting' and 'being affected', as 'agent' and 'patient' (SB).

57. I am here translating the text printed by the Aldine editor. To translate the

MSS text as given by Vitelli would have meant presenting the reader with the nonsensical 'a large amount of fire by a small amount, for instance'.

58. The Greek is *pephuken oude.*

59. The phrases 'for if like' and 'by like' translate words inserted into the text by the Aldine editor and by Vitelli.

60. The Greek here is *tên drastêrion ekhein dunamin.* Philoponus sometimes seems to give *dran* a more specific use than *poiein*, more usually used for acting. The phrase *drastikon aition* can be used of efficient cause: cf. *in GC* 53,18, where *drastikê* is contrasted to *pathêtikê*, capable of being affected. Mourelatos points out that Philoponus uses it to point out things with specifically 'efficacious power': 'Aristotle's rationalist account of qualitative interaction', *Phronesis* 29 (1984), 1-16, p. 7. The distinction is perhaps that while a body as a whole is described as either active or passive in a change, while *drastêrios dunamis* is often used to designate the specifically active element, such as its heat. This usage comes up again in later chapters to distinguish the primary qualities to which causation can be reduced from other contraries that change into each other (SB).

61. The Greek terms *poiêtikon* and *pathêtikon* could mean either 'active' or 'capable of acting', 'affected' or 'capable of being affected'. But the context requires a contrast between the fact that while both alike have the same capacities, one is actually active, the other affected (SB).

62. Literally, 'into itself'. [On criticisms of this model of change, see Introduction and 146,31ff. below (SB).]

63. This argument seems not to follow: at the very least, a lot of work is extracted from the *pantôs*, 'in every way'. Showing that things effect one kind of change does not show that they will be destroyed: this would undo the argument for an eternal fifth element. Philoponus seems to think that having the capacity to be changed is the critical move, presumably because things that undergo change have a form-matter structure and are thus susceptible to destruction. But if the fifth element only has *topikê hulê* – 'matter' for local change, however that is to be understood – it would not be susceptible to destruction. Aristotle says, not that everything will *destroy* itself, but that everything will *move* itself. In other words, he seems not to be worried about the potential for self-change leading to self-destruction, but that making everything self-identical capable of self-motion will do away with the argument for an eternal unmoved mover (SB).

64. Book 8.4ff.

65. The verb here is *existêmi*, a term used as definitive of the idea of change; here the idea of something being dislodged from a former state is introduced as a contrast to change *per accidens.* On the idea that alteration involves introducing a new state, see also nn. 47 and 95 (SB).

66. The word translated 'happens' here is a verb whose participle is represented by *accidens* in the phrase translated *per accidens.*

67. The word translated *simpliciter* here means 'without qualification': presumably he is saying that what is being asserted is not 'Every body is acted upon by some body' but 'Every body that is acted upon is acted upon by some body'. He would be anxious to ensure that nothing was said to prejudice the immunity of the celestial bodies from affection of any kind.

68. See ch. 1.5 in the companion volume.

69. The Greek term is *homoüla* (SB).

70. Philoponus takes the three-dimensional extension, an abstract notion shorn of any particular quantitative determination, to be prime matter. On this passage see Frans de Haas, *John Philoponus' New Definition of Prime Matter*, pp. 155-6. The idea that qualities and quantities are realized within a certain latitude, a range of possible values, becomes very important in Philoponus' account of change (SB).

71. *epiteinomenôn tinôn poiotêtôn ê aniemenôn*. The Greek terms *epitasis* and *anesis*, from which these participles are derived, have a subsequent history in Latin tradition and are often translated as 'intension' and 'remission'. The idea that qualities are realized within a certain *platos* or range of possible ratios of a mixture is translated into the Latin tradition as *latitudo* and hence is often translated as 'latitude'. See Nelly Tsouyopoulos, 'Die Entstehung physikalischer Terminologie aus der neoplatonischen Metaphysik', *Archiv für Begriffsgeschichte* 13 (1969), 7-33; R.B. Todd, 'Some concepts in physical theory in John Philoponus' Aristotelian commentaries', *Archiv für Begriffsgeschichte* 24 (1980), 151-70, p. 163; Katerina Ierodiakonou, 'Aspasius on perfect and imperfect virtues', in R.W. Sharples (ed.), *Aspasius*, Berlin 1999. These ideas become very important in Philoponus' account: see Introduction, also n.72 and 142 below, and ch. 1.10 below (SB).

72. The idea that qualities admit of degree is attuned by Philoponus to the view that qualities supervene on mixtures, developed in 1.10, by the idea that the mixture of the four elements required to constitute a substance like flesh is not a precise ratio but admits of a range. Thus it is true both that changes in the quality vary with changes in the mixture although not every change in the mixture need affect every quality, but also that different qualities supervening on different mixtures could all be qualities of the same substance like flesh. All these are possible because there is a range of mixtures possible for a substance and slight differences of the mixture could still form the same substance. See previous note and Introduction (SB).

73. The Greek here is *prosekhôs*, the term Williams had previously been translating as 'proximate'. The point is not about temporality but about the absence of a mediary (SB).

74. Ch. 5ff.

75. This sentence, *pan gar melan sunkritikon opseôs kai pan leukon diakritikon*, is very close to one from Theophrastus' *De Sensu* 86 cited by LSJ s.v. voce *sunkritikos*, namely, *to men melan sunkritikon, to de leukon diakritikon*. [The terminology derives from Plato's *Timaeus* 67D-E. See also Philoponus *in DA* 188,1; 349,7; 439,18 (SB).]

76. According to Vitelli's apparatus no manuscript reads 'water' here; some MSS have 'fire', others omit the clause about the cold in the water altogether. If the clause is to be included, 'becomes the hot which is in this water' seems to make more sense than '... in this fire.' This note is due to David Furley.

77. *Metaph*. 4.3, 1070a13-26; 8.4, 1044b21-4 (RRKS).

78. Vitelli prints the words *ean ara* between commas in the Greek text corresponding to this point. The Aldine editor, who, presumably, like me, despaired of making sense of these words, omitted them, and what I have written may here be regarded as the translation of his text.

79. Sambursky drew attention to this passage in his argument that 'suitability' had a technical meaning: 'Conceptual developments and modes of explanation in later Greek scientific thought', 61-78 in A.C. Crombie (ed.), *Scientific Change*, p. 73. See Introduction (SB).

80. It seems more helpful to translate *haplôs*, which I usually render *simpliciter*, by 'without qualification' at this point.

81. Book 1.7, 191a4.

82. The word I have been translating 'immediately' in the previous pages was *prosekhôs*, which is adverb from *prosekhês*, meaning 'next'. The word here, *amesôs*, has an even better right to be so translated.

83. sc. things that cause change and are changed.

84. Ch. 3, 1b10: e.g. if man is said of Socrates as of a subject, animal, which is said of man, will be said also of Socrates. [The term *hupokeimenon*, literally 'underlying thing', is generally translated 'substratum', but can also mean 'subject', i.e. the subject of predication (SB).]

85. The *thumos*, a term from Plato's tripartite division of the soul rather than Aristotle's theory (SB).

86. Vitelli has added *en* here, surely rightly.

87. Philoponus no doubt has in mind the familiar example of the hand which uses a stick to move a stone.

88. Aristotle's term for a word used equivocally: see n. 21. Thanks to David Robertson for pointing out an error here (SB).

89. Categories are not, in Aristotle's view, kinds *of* anything except being, and being is not a genus.

90. Vitelli follows the Aldine editor in printing *hulên <tên> autên ousan* instead of the MSS' *hulên autên ousan*.

91. The MSS need a particle at this point, as Vitelli remarks. Curiously, however he does not print in the text the *oun* which he suggests in his apparatus. I have made good the deficiency.

92. The root meaning of *kata metaphoran* is 'by transferring', whereas *katakhrêstikôs* means a misapplication.

93. We treat sickness for the sake of health: the question is whether health can be said to be the cause of becoming healthy.

94. The Greek is *hexis*, which might mean 'state' or 'disposition'. [On the contrast between changing something to a new state and actualizing a state that is already present, see Introduction and n. 47 above. To realize a child's potential to do geometry requires an agent leading the child to a new state or disposition. But to realize the potential in a sleeping geometer requires no new state, merely the removal of the impediment to the actualization of a state or disposition that is already present. Philoponus is pairing the first kind of causation to that of efficient causes, the latter to that of final causes (SB).]

95. The word I have translated 'efficient' here is the same as has been translated 'capable of acting' or as 'agent' above. [See nn. 47-9 and Introduction, above. Philoponus seems here to be stressing the idea that to act, *poiein*, is an inherently temporal notion, to explain Aristotle's reluctance to call the unmoved mover an efficient cause (SB).]

96. This not very flattering metaphor for the divine activity is based, presumably, on the consideration that God does not so much do things as get others to do them. It is used by other authors.

97. Williams translates *pathêtikon* here as 'passive' rather than 'capable of being affected', to emphasize the connection to *apathes*, here translated 'impassive'. Williams sometimes translates the terms for things which act and are affected as 'agent' and 'patient' (SB).

98. The Greek word translated 'passage' is *poros*, i.e. 'pore'. It is a 'way through', precisely what is missing when one is faced with an *aporia*, an inability to see one's way through a particular difficulty.

99. These words and the preceding lacuna correspond to the Greek text printed by Vitelli. It is not at all clear what has happened to the text.

100. The Greek is *sullogismos*; the argument form is a conditional, not the Aristotelian syllogism. The first version is of the form 'if p, then q; p; therefore q', where p = 'there is no void', q = 'there is no movement'. Philoponus reformulates this in *modus tollens* in positive rather than negative terms (p = there is movement, q = there is void): 'if p, then q; not q; therefore not p.' Democritus rejects the assumption 'not q' in the second formulation, arguing instead by *modus ponens* 'p; therefore q' (SB).

101. *ta enargê ek tôn aphanôn anairei*: the terminology stresses the perceptual nature of the evidence. Philoponus is commending Democritus' procedure of positing imperceptible entities only when necessary to explain observed phenomena and not controverting the evidence of the senses (SB).

102. The Greek word *arkhas*, which I have here translated 'fundamental constituents' is frequently translated 'principles', but this suggests something propositional which is not appropriate here.

103. Reading *autôn* rather than *autois* here at Furley's suggestion (SB).

104. Here Williams is translating a slightly different text used by Philoponus than that of the modern text of Aristotle. Instead of *hodôi de malista kai peri pantôn heni logôi diôrikasi Leukippos kai Dêmokritos*, which Williams rendered in the Clarendon as 'the most systematic theory, however, which covers all the phenomena with a single explanation, is that of Leucippus and Democritus', Philoponus' lemma reads *hodôi de mallon kai peri pantôn diôrikasi Leukippos kai Dêmokritos*. I have changed the Clarendon translation of the next lemma accordingly to preserve sense (SB).

105. Here I am translating *arkhais* as 'starting points' (cf. n. 102 above).

106. Or 'does not exist' or 'is non-existent': the Greek is *ouk on*, which allows for the various renditions that Philoponus considers (SB).

107. The atomists divided space into that which is occupied by atoms, which they called *to plêres*, 'the full' or the *plenum*, and that which is not so occupied, which they called *to kenon*, 'the empty' or the *vacuum* or void.

108. This argument is attributed to Melissus at *Physics* 4.6, 213b12 and 214a27 (SB).

109. Book 4.6, 213a12ff. [Here Aristotle is not reporting his own view. Philoponus does come to endorse the view that space is void in its own nature but always filled with body in his *Corollaries on Place and Void*, translated by David Furley, London and Ithaca N.Y. 1991 in this series, pp. 567,31-4; 582,20; see David Sedley, 'Philoponus' conception of space', in Sorabji (ed.), *Philoponus and the Rejection of Aristotelian Science*, London and Ithaca N.Y. 1987, 140-53 (SB).]

110. The term translated 'artificiality', *peplasmenôs*, describes the difficulty of providing a reason why divisibility is selective, occuring here rather than there. The vivid image of intrusion by a party of revellers in the verb *eiskômasei* emphasizes the disruptiveness of the hypothesis of void (SB).

111. The word translated 'need' (*dein*) occurs twice in this sentence, but it seems impossible to make sense of it at its first occurrence. I have therefore assumed that it should be deleted. [Furley suggests reading *mêden* instead of *mê dein* for the second occurrence, which would solve the problem (SB).]

112. Aristotle's point is that coming-to-be is produced by clustering and that alteration happens at points where the atoms have come in contact but are not truly unified. All this is meant to show how the atomists try to avoid Parmenidean arguments against change (SB).

113. Literally, 'to a lesser extent follow from and are the consequence of'.

114. *Timaeus* 54Cff. The four simple bodies are formed from solids composed of different triangular surfaces.

115. *Timaeus* 56B.

116. Presumably Philoponus feels a need to explain why there was a big jump from the octahedron to the eicosahedron.

117. The modern text here reads *hai geneseis kai hai diakriseis*, which Williams translates 'comings to be and segregations'. But Philoponus' comment indicates that he reads the text to be claiming that there are two means of comings to be. He tries to explain how this might be the case although it seems that he would have preferred the modern text (SB).

118. The MSS of Philoponus have *adiaireton*, undivided, here, while the MSS of Aristotle (with one exception) have *diaireton*, which is clearly what Philoponus read in Aristotle. This footnote is due to David Furley (SB).

119. Book 1.2, 315b31.

120. See companion volume, ch. 1.2.

121. sc. the mathematicians.

122. sc. the atomists.

123. The division in question is division into two *equal* halves.

124. The Greek word translated 'self-awareness' is *sunaisthêsis*, literally 'co-perception'. Perhaps this word is an ancestor of 'consciousness'. [See, e.g. Wolfgang Bernard, 'Philoponus on self-awareness', in Richard Sorabji (ed.), *Philoponus and the Rejection of Aristotelian Science*, London and Ithaca N.Y. 1987, 154-63 (SB).]

125. sc. in the interstices between the constituent atoms.

126. This word, and the words inserted in the following sentence represent words that the Aldine editor supplies for what appears to be a lacuna in the text, which Vitelli represents as such. The words supplied are printed by Vitelli in the apparatus as follows: *hapsasthai. ei men oun en tôi kenôi khôrêsai, ouk aisthêsetai. hên* (sic) *de tês atomou*.

127. sc. the atoms.

128. 152ff. [See Myles Burnyeat, *The Theaetetus of Plato*, Hackett 1990 (SB).]

129. sc. that of atomism.

130. The subject of both verbs is unclear: Williams reads the first to be plural *elegon* rather than *elegen* of the MSS. Another suggestion by Furley is to repunctuate the sentence and read both as plural, i.e. 'But they said <the atoms> are all heavy, as they also said that they also move downwards'. On whether the early atomists attributed weight to the atoms see Denis O'Brien, *Theories of Weight in the Ancient World, Vol. 1. Democritus: Weight and Size*, Leiden 1981, 153-61. Another possibility is that the second 'he said' refers back to Plato: cf. Simplicius *in Cael.* 268,32ff., and O'Brien p. 161 (SB).

131. sc. *On the Heavens*.

132. The thought, which is not particularly transparent, seems to be that, if the ratio of the amount of void in A to the amount of void in B is greater than the ratio of the amount of plenum in A to the amount of plenum in B, A is lighter than B.

133. These are words inserted by the Aldine editor and printed by Vitelli.

134. This phrase and the same phrase at its next occurrence are my own insertions.

135. The word I have translated 'branches' would more literally be translated 'segments', but modern philosophers are more used to examining the range of possibilities by using the device of trees and branches rather than division into segments.

136. The question how a composite could have a quality not possessed by any of its components had been discussed earlier at *in GC* 1.2, 31,7, with regard to the question how things not possessing size could give rise to something that has magnitude (SB).

137. If nature consisted of impassive atoms, it would not be in the nature of anything to be affected, and therefore not in the nature of anything to perceive (RRKS).

138. Something belongs to Aristotle's category of relation if it is what it is only in relation to something else: e.g. a wife is necessarily *someone's* wife.

139. Vitelli, no doubt following the MSS, prints *oude* in front of the words translated 'in relation to perceptions', but this seems to make the opposite point to the one which Philoponus is clearly concerned to make. I therefore follow the Aldine editor in omitting *oude*.

140. 152ff.

141. See e.g. Galen, *Quod animi mores*. Philoponus discusses the medical view more extensively at *in DA* 50,16ff., where he makes the criticism that if passions result from or were products of the mixing, they would not be susceptible to the influence of philosophy. In the *De Anima* commentary, a contrast is drawn between qualities which are products of mixture, and powers which supervene on a suitable mixture but do not follow it of necessity: *in DA* 50,16-53,8; 141,22-9; 439,35-440,3. See Todd, 'Philosophy and medicine in John Philoponus' commentary on Aristotle's *De Anima*', *Dumbarton Oaks Papers* 38 (1984), 103-10, p. 110; also 'Galenic medical ideas in the Greek Aristotelian commentators', *Symbolae Osloenses* 52 (1976), 117-34. Philoponus' *in Phys.* compares the harmony imposed on the lyre by the craftsman to the 'lives' imposed from outside (*in Phys.* 191,18-24). The verb used here, *epiginetai*, translated 'supervene', is also meant to indicate that the forms are not produced by the mixture: it does not exactly match the meaning of supervenience in modern philosophical debates. For an excellent overview of recent debates and the parallels to ancient discussions, see Victor Caston, 'Epiphenomenalisms, ancient and modern', *Philosophical Review* 106 (1997), 309-63. In her 'Aspasius on perfect and imperfect virtues', Katerina Ierodiakonou points out that the terms used in the discussion of degrees of qualities, *epitasis* and *anesis*, like that of attunement, have their origin in music. See also nn. 71 and 72 above, and Introduction. I am grateful to Richard Sorabji and Victor Caston for references and clarification (SB).

142. Todd finds precedent for this expression, *tês holês dêmiourgias*, in Proclus *in Tim.* 3,53.6 and Damascius *in Phaed.* 1.3, where it means the intelligible realm in contrast to the visible cosmos: 'Some concepts in physical theory in John Philoponus' Aristotelian commentaries', p. 163 n. 54. The same phrase can be found in Philoponus *in DA* 52,15-16. On the idea that the intelligible realm is created, see R. Sorabji, *Time, Creation and the Continuum*, pp. 250-2; Jean Pépin, 'Recherches sur le sens et les origines de l'expression *caelum caeli* dans le livre XII des *Confessions* de S. Augustin', *Ex Platonicorum Persona*, Amsterdam 1977 (SB).

143. I suspect that this is an allusion to a view about the acquisition of form by matter which is still extant in late medieval philosophy, namely, that matter has to be organised in such a way that it is 'disposed' to receive a form, although these 'dispositions' in the matter are not of themselves sufficient to cause the matter to acquire the form. This theory has particular importance when the form in question is a soul. [The difficulty for an Aristotelian commentator is Aristotle *DA* 1.1, 403a19-22, which, without going as far as sufficiency, goes a good part of the way (RRKS).]

144. Philoponus seems to think that this problem applies to any account in which the qualities that can inhere are determined by the matter of the whole body, whether they produce qualities or the qualities supervene on the matter (SB).

145. Literally, 'differentiae' (*diaphorai*).

146. The sense in which the word thus rendered (*seiraiou*) must be understood here seems to be ignored by LSJ. It is probably connected with *Seirios*, the dog-star.

147. The word translated 'point-like' here is *atomos*, which can mean 'indivisible' (or, in other contexts,' atom'). The claim is that qualities can admit of degrees, since one can distinguish different ends of the range covered by a given quality-description. Philoponus uses the term for a quality rather than an object: the problem arises in a critique of atomism and its inability to explain the inherence of multiple qualities in an indivisible matter. On 'range', see nn. 71-2 above (SB).

148. Sambursky draws attention to this passage in 'Conceptual developments and modes of explanation in later Greek scientific thought' (SB).

149. This translates *atomou*, which here functions as an adjective. [See n. 147 above. Philoponus seems to be suggesting that the conditions for the supervenience of qualities have precise and not vague boundaries (SB).]

150. The MSS read at this point *to kuriôs on to auto to de pantelôs êmeiphthai*, which Vitelli describes in the apparatus as '*corrupta*', placing an obelus in the text after *auto*.

151. Thought of as a very fine substance ('subtle matter'), as in 'animal spirits'. [The term *pneuma*, while it does appear in Aristotle in some substances like wine and semen, is widely used in Stoic and medical sources. Aristotle uses *pneuma* to explain the white colour of froth in *GA* 2.2, 736a1-22; 736b37 (SB).]

152. Philoponus has been talking about 'parts' of atoms since 168.22ff.: not separable parts, but parts distinguishable in thought (as in Epicurean atomism). This note is due to David Furley.

153. sc. the atoms.

154. 'Primary and fundamental constituents' represents *prôtai arkhai*; see n. 102 above.

155. cf. Plato, *Phaedo* 71E.

156. The phrase 'by classes' translates *kata summorian*: a *summoria* was a taxation group of Athenian citizens, but the phrase had probably become a dead metaphor by the time of Philoponus.

157. Here I have translated *kata summorian* 'different classes having different classes'.

158. See n. 135 above.

159. The Greek *adiaireton* is ambiguous as between 'indivisible' and 'undivided'.

160. See n. 102 above.

161. Philoponus has argued that if atoms are in fact undivided but could be so, there must be some basis for internal distinctions, and this could be used to allocate different capacities. This argument seems to understand 'indivisible in potential' as meaning that the atoms are not divided into parts even potentially. But Philoponus goes on to argue as if he thinks this means that an atom has its potentialities undividedly, i.e. that no potentiality could belong to only part of an atom and hence that no potential can be affected without the others also being changed. These would only be equivalent if it is assumed that different potentialities must inhere in different parts to be affected independently of one another. The earlier discussion of 167,27ff. may be the background assumed here. Philoponus' own doctrine of 'range' or 'latitude' aims to show how affections belonging simultaneously to the same whole may be affected separately (SB).

162. This seems to refer back to the argument about atoms having different affections at 167,27.

163. In ch. 2, 316a4.

164. See n. 102 above.

165. The Greek for these two questions differs only in one letter: *posa tauta* and *poia tauta*.

166. The Greek could read either thus, as Philoponus clearly reads it, or, as Williams translated in the Clarendon, 'they would act and be acted upon in coming in contact with one another' (SB).

167. The Aldine editor's addition of *ara* is accepted by Vitelli.

168. The second 'through' here is an addition by Mugler, the Budé editor, which seems essential if sense is to be made of the text. See Philoponus' own comments, 178,29-33 below.

169. sc. by coming into contact with it at this point.

170. Book 4.6-9.

171. The modern text has *ei plêrês hekatos*, which Williams translates as 'if they are each of them full' in the Clarendon edition. What Philoponus quotes at 179,7 is *ei men kai peplêrômenôn tôn porôn* (SB).

172. Above, 178,10ff.

173. The use of the term for contact in the plural has precedent in the previous passage at 178,17. While Empedocles (and Plato) talk of vision as an *opsis* extending from the eye – a term later used in geometrical optics for the multiple 'rays' between an object and the eye – they seem to use the term in the singular: Aristotle *Sens.* 2, 437b14; *GA* 779b15ff.; Theophrastus *Sens.* 7,5; 17,4; 19,4 (Plato *Tim.* 64D, 67C-E; *Rep.* 507D). See nn. 176-9 below (SB).

174. sc. in three dimensions.

175. *mataion*. The term suggests the Aristotelian dictum that nature does nothing in vain (SB).

176. *kata moria* in Greek, which might mean 'cut into pieces', or else that it is cut at particular points rather than others. See Simplicius *in Phys.* 693,10ff., who defends Aristotle against the view that the sun's rays pass through pores in a medium; also *in DA* 134,12. Some had argued that there must be void passageways allowing rays to pass through, or full jars of water should overflow when light goes through. The advocates of passageways 'dividing up' a continuous medium argue that this is why rays are reflected at some points while others pass through at other points. A very similar argument seems to be made here with regard to the 'visual ray' hypothesis for sight: that like the 'ray' theory of light, it would require passageways scattered through a medium (SB).

177. Philoponus also raises this criticism at *in DA* 325,26. The idea that vision occurs by multiple discrete rays connecting the eye and the object may owe something to the geometrical model of optics rather than the early use of *opsis* for the faculty of sight: see my 'Euclid and the sceptic: a paper on vision, doubt, geometry, light and drunkenness', *Phronesis* 43 (1998), 176-96 (SB).

178. This is expressed more fully at *in DA* 325,19. The argument is that if we see a quarter of the heavens when we look, if visual rays were a body they would either need to pass through the intervening body – the air – or else to divide it. But body can't pass through body. And if the visual rays were divided on passing through the intervening air, there would need to be empty passageways in the air; moreover we wouldn't see objects as a continuous whole (SB).

179. Imagine two people side by side looking at the same tree: each will see a tree by a cone-shaped array of visual rays centred on their eye with the base at the tree, so the cone-shaped arrays will partially intersect. If the rays are bodies, then, they would have to pass through one another (SB).

180. The added words are due to Vitelli, who is surely right here. [Furley suggests inserting <*kai kenon*> instead of <*kai ou plêres*>, 'so the void is both void and full' (SB).]

181. Again following Vitelli, who prints *sôma ti* in place of *sômati*.

182. Here Philoponus seems to be referring to a 'microvoid' theory. Some scientists after Aristotle held that small void spaces exist while large void spaces can't, including Erasistratus of Ceos, an important medical researcher of the third century BC, and Hero of Alexander, an engineer from the first century AD. Since Hermann Diels' article, 'Über das physikalische System des Straton', *Sitzungsberichte der Prüssischen Akademie der Wissenschaften* (1893), 101-27, there has been much discussion about whether a microvoid theory can correctly be attributed to the Peripatetic Strato of Lampsacus: see esp. H.B. Gottschalk, 'Strato of Lampsacus: some texts', *Proceedings of the Leeds Philosophical and Literary Society*, 11 (1965), 95-182; David Furley, 'Strato's theory of void', in *Cosmic Problems: Essays on Greek and Roman Philosophy of Nature*, Cambridge 1989, 149-60. Simplicius' report about the passage of light through pores, mentioned in n. 176 above, is central to this debate (SB).

183. This argument ignores the difference between what void is in its own nature and whether void ever exists actually deprived of body: there might be other reasons why large void spaces never actually exist. For Philoponus' recognition of the distinction between the logical and the physical impossibility of void, see David Sedley, 'Philoponus' theory of space', in Sorabji (ed.), *Philoponus and the Rejection of Aristotelian Science*, London and Ithaca N.Y. 1987, 140-63 (SB).

184. Book 1.8, 191b33. It is not at all clear what Aristotle means by this sentence, and it is hardly possible to interpret 'this nature' as a reference to the distinction between potentiality and actuality.

185. It isn't clear who these interlocuters might be. Early Peripatetic texts certainly do make use of passageways in connection with sweat: see *Problemata* 2,8.17,20,23,41; Theophrastus *de Sudore* 25-7 (SB).

186. It is not so much that the veins themselves are in some special way more easily affected but less easily affected *simpliciter*, as that the minerals are more easily affected in respect of their veins but less easily *simpliciter* because of the disparity.

187. Here, the translation departs from the Clarendon because Philoponus seems to quote a slightly different text than the modern (SB).

188. This refers back to the previous chapter, 163,28ff. The term for 'contact' in the plural does not seem to occur in the *Timaeus* passage that discusses the composition of solid bodies from plane figures (SB).

189. Vitelli indicates a lacuna at this point.

190. See the previous volume, Book 1.2, commenting on 315b35: there Philoponus remarks that '*Rhusmos*, *tropê* and *diathêkê* belong to the dialect of Abdera, and they are used by Democritus, who calls shape *rhusmos*, position *tropê* ... and order *diathêkê*', *in GC* 26,8-11, Williams' translation. *diathêkê* and *diathêgê* are variant MS readings (SB).

191. i.e. in their theory there is no room for such a phenomenon as coming-to-be, as Aristotle conceives it.

192. Vitelli indicates a lacuna at this point.

193. I translate *mixis* as 'mixing' and *krasis* as 'mingling', not that the meaning of the two English words does much to illuminate Aristotle's use of the two words here.

194. Vitelli prints *prôton ton* (sic) *problêmatôn*, where *ton* must be a mistake for *tôn*.

195. Philoponus has slipped here. He should have said 'they would be no more mixed after mixing than before mixing' or 'they would be no less mixed before mixing than after mixing'. Aristotle says, 'now they are no more mixed than they were before' (327b1-2).

196. *homoiôs ekhein.*

197. i.e. watered down. [For the earlier use of suitability, see Introduction and n. 79 above (SB).]

198. *eilikrinôs*, i.e. in the way that is distinctive for the disposition, in contrast to restrained functioning: see n. 307 and De Haas, 'Mixture in Philoponus. An encounter with a third kind of potentiality' (SB).

199. The Greek word which I have translated 'mixture' is *kramati*: its literal meaning has the special sense of the result of mixing water with wine.

200. The Greek is *kekolasmenôs*, meaning literally 'chastised'. Perhaps the metaphor is drawn from the use of the word to describe a horse which has been restrained from plunging ahead with its full vigour. Other forms of the same word occur at 188,25; 192,12; 198,27 and 202,22. [The idea that in a mixture the components restrain each other's functioning, in contrast to the pure activity, is discussed in De Haas' paper 'Mixture in Philoponus. An encounter with a third kind of potentiality' (SB)].

201. *De Anima* 1.2, 403b21.

202. The Greek, *to mikton*, mean either that which is mixed, i.e. the mixture, as Williams translates it in the Clarendon, or that which can be mixed, i.e. the ingredient. That Philoponus reads it the latter way is clear from his discussion (SB).

203. Although the relevant part of Aristotle's text is provided in full in this translation, Philoponus' remarks seems to refer to the phrase 'and what it is that is capable of being mixed and to which of the things that are it can be attributed', which is all that is written out in the manuscript of the commentary. The pleonasm is that attribution can only be made to the things capable of being mixed (SB).

204. In the Greek the single word *alla.*

205. Philoponus and some MSS read *oute* at 327b10, where Joachim prints *oude*, which in my Clarendon translation I rendered by italicizing 'not'.

206. Presumably, this example involves a stricter sense of 'being in one and the same substratum, because both affections are affections of a body, whereas whiteness and knowledge were affections, one of the soul or mind, the other of the body.

207. The Greek here is *stratiôtôn potamou.* [This is commonly thought to be a kind of water lettuce. Alexander of Aphrodisias cites a Stoic demonstration of separating water from wine by dipping in an oil-soaked sponge (SB).]

208. There are different criticisms of Anaxagoras that might be referred to here. One is that mentioned earlier, that because an affection or property of a substance does not subsist independently, these cannot be said to be mixed. As Philoponus expresses it, the counterexample might merely be challenging the idea that *all* things are mixed or that the concept of mixture could adequately account for the relationship of affections to substances. But the claim that in mixture things are only themselves in potential might also seem to preclude universal mixture on the grounds that nothing would ever have its true form. On the idea of *minima naturalia*, the idea that Aristotle rejects Anaxagorean universal mixture for the quite different reason that parts of organisms cannot be indefinitely small in size, see Michael Wolff, *Fallgesetz und Massebegriff*, Berlin 1971, pp. 130-46; R.B. Todd, 'Some concepts in physical theory in John Philoponus' Aristotelian commentaries', 151-70, *Archiv für Begriffsgeschichte* 24 (1980), p. 161 (SB).

209. The Greek term *diairêteon* can mean 'distinguished', but Philoponus tries to give a meaning to its more literal sense of divided into parts (SB).

210. The term *homoeomer* refers to a substance whose parts are the same: every part of flesh or blood is also flesh or blood. This is in contrast to anhomoeomerous parts: no part of a hand is itself a hand (SB).

211. The phrase 'one which is composed of six ingredients' translates, literally, the Greek *hexaeidos*. Presumably this was the equivalent of a 'brand name'.

212. Book 1.2, 185b10, and elsewhere.

213. What Philoponus actually says is that each of the two assumptions is used to refute each of the two theses, but it is clear from what he says at 194,4-5 that he uses one assumption to refute one thesis and the other to refute the other.

214. Following a suggestion by Richard Sorabji (SB).

215. If the continuous is divisible to infinity, mixture can't be the same as putting together of smallest parts (because there would be no smallest parts). If mixture is different from juxtaposition of smallest parts, mixture won't be relative to perception but there must be no point at which it resolves into a juxtaposition of smallest parts, so there must be infinite divisibility. The atomists wouldn't accept either assumption, of course.

The middle argument claims that an individual atom doesn't retain the character of an element, i.e. isn't fire by itself but only in a mass. The atomists would want to say that the character of the component atoms is unaffected, so have an initial difficulty explaining the fact that a composite doesn't always display the qualities its component atoms would display if they were unmixed. They solve this difficulty by distinguishing between the qualities of atoms themselves and qualities that are apparent to observers. Thus it is reasonable that, say, fire atoms in a cluster might display a character that would not be evident in a mixture containing fire atoms (SB).

216. The point is clearer in Greek, since the terms translated 'every' are of different genders, corresponding to the genders of barley and wheat (SB).

217. This seems to be the only indication that Philoponus has noticed the carelessness of Aristotle's formulation of this possibility: it is, of course not possible that every grain, or whatever, should be in contact with *every* grain, but the situation he wishes to describe is the one where every grain of the one sort is alongside *some* grain or other of the other.

218. The point is that a number of particles, say five, of some very fine powder may be together in the mixture, so that the arrangement is ABABAAAAAB-BABAA, rather than the perfect ABABABABABABABA, and the different kinds of component still be indiscernible by people with normal vision.

219. Philoponus seems to allow an extended sense of homoeomereity to an atomist, i.e. that at the perceptible level the substance is perceptibly indistinguishable when the atoms are evenly distributed in a juxtaposition (SB).

220. Gisela Striker pointed out that *logos* here refers to definition, since homoemoeries are characterized by having parts same in nature as the whole (SB).

221. Actually, the main clause on which both the 'neither' and the 'nor' clauses depend is the clause translated 'It is right, clearly, to say that things have been mixed'.

222. The phrase 'the earlier things he has discussed' translates the Greek *tôn prôtôn* – 'the first things'. It seems clear that the things he has in mind are the topics of the previous chapters: coming-to-be, alteration, growth, contact, acting and being affected.

223. These words have to be deleted if we are to make sense of his description of 'being mixed' (*mignumenê*) as 'used once to apply to both subjects' (*sullêptikôi tropôi eirêmenon*).

224. Reading *mignumenê* with two MSS and the Aldine editor, rather than *mignutai* which is printed by Vitelli, following two other MSS.

225. Aristotle omits words that are to be understood as applying to both subjects. Philoponus first suggests that the Greek be filled out understanding both 'produce health' and 'by being mixed with bodies' with each subject. The second suggestion is that only 'produce health' is to be understood with both, 'by being mixed with bodies' only with 'medical skill' (SB).

226. In Greek *logos*.

227. The Greek word translated 'easily bounded' is *euoristos*, that translated 'easily divided' *eudiairetos*, and that translated 'liquid' *hugros* (which elsewhere requires the translation 'wet' or 'moist'). [Philoponus' lemma has only *diareta*, 'divided' (SB).]

228. On the doctrine of *minima naturalia*, the view elaborated by later interpreters on a remark of Aristotle at *Physics* 1.4, 188a1 that there is a certain minimum quantity below which certain substances cannot be realized, see Todd, 'Some concepts in physical theory in John Philoponus' Aristotelian commentaries', p. 16; De Haas, *John Philoponus' New Definition of Prime Matter*, p. 157. Strictly Aristotle says this of substances occuring in living bodies, but here Philoponus seems to see it as having wider application, as he does in *Contra Proclum* 431,12-16: see De Haas, p. 161 and Introduction, above. Richard Sorabji points to *Physics* 7.5, 250a9-25, b4-7 for other restrictions to quantity that don't depend on the idea of a *minima naturalia* (SB).

229. *epi polu*.

230. *amblunêtai*, a term originally used of the dulling of sharp instruments, but used by transference of wine losing its force. Todd suggests that the medical tradition may be a source for Philoponus' idea of a *platos*, despite the lack of evidence for the pair *anesis/epitasis* in Galen's pharmacology: 'Some concepts in physical theory in John Philoponus' Aristotelian commentaries', pp. 168-9. The term *amblunetai* is used by Galen frequently and sometimes contrasted with *epitasis*, e.g. *ad Glauconem de med. meth.* 11.112,16; *de simpl. med. temp.* 11.722,3, and used of the tempering of powers in mixture (SB).

231. *huperbolas kolazei*. Compare Aristotle's remark at *GC* 2.7, 334b8ff., where 'excesses' are said to be destroyed rather than restrained. See De Haas, 'Mixture in Philoponus. An encounter with a third kind of potentiality' (SB).

232. Philoponus' lemma has the words *ek toutôn* which are not in the MSS of Aristotle.

233. i.e. in the sublunary world.

234. It isn't clear that is the point Aristotle has in mind: on Philoponus' concern to show that the text includes the heavenly bodies, see the previous volume and chapter 1.6 n. 18 above (SB).

235. Vitelli's apparatus at this point reads *immo ta mikta*, surely rightly. The words 'the things which act and are affected' are surely a slip for 'the things which can be mixed'. Philoponus surely began his sentence meaning to pick out the class of things which can be mixed as a subclass of the things which act and are affected.

236. The translation of this passage owes much to Gisela Striker (SB).

237. Suppose a piece of gold were divided into ten equal parts, each weighing six ounces. If the parts were put together again the whole would be expected to weigh sixty ounces. Philoponus is claiming that it would be observed to weigh more than sixty ounces.

238. *sumpnei mallon kai zôoi to holon*. One MS has *sumplei*, 'sail together with'. The idea that mere addition of like parts should increase the powers of the whole disproportionately might seem to be an example of 'unmechanistic' reasoning or an antiempirical approach. But there are cases where this does seem to be observed: the intenser burning of a larger fire, or the greater inertial force required to move a large mass. For the use of 'lives' as a synonym for powers, see the next chapter, 206,13 (SB).

239. See n. 227 above. [The term 'easily bounded' has the sense of something fluid here, i.e. conformable to a container (SB).]

240. Some of the MSS here read *pampan ê êrema*, which is what Vitelli prints. Others omit the *ê*, which makes sense, and is the text printed by Joachim. Vitelli's text, which has some support from MSS of Aristotle, seems to make no sense at all.

241. cf. n. 200.

242. Philoponus has *eiê* here, where the modern text of Aristotle reads *ê*.

243. Vitelli marks a lacuna at this point. [Gisela Striker suggests that this is probably referring to the atomist argument (SB).]

244. Philoponus has *eiê* here, where the modern text of Aristotle reads *ê* .

245. Aristotle's definitions of 'synonymous' and 'homonymous' in Chapter 1 of the *Categories* are as follows: 'When things have only a name in common and the definition of being which corresponds to the name is different, they are called homonymous' (1a1-2), and 'When things have the name in common and the definition of being which corresponds to the name is the same, they are called synonymous' (1a6-7). However, he does not always conform to his own prescription for the use of these words, and this is presumably Philoponus' reason for calling his usage here 'habitual with him'. [Mugler in the Budé edition of *GC* refers to *Physics* 7.4, 249a21-5 here (SB).]

246. See n. 200 above.

247. 'Full strength' translates *akrotêta*. The term is related to the idea of something being at its peak and was used in conjunction with the idea that there are degrees for qualities (SB).

248. In the Clarendon Williams supplies a pronoun and translates *diaporan autôn* as 'differentiates it from them', i.e. alteration from coming-to-be and perishing. Philoponus reads the text a little differently, as he explains, by supplying 'each', referring to alteration, coming-to-be and perishing (SB).

249. Vitelli prints a lacuna at the beginning of the sentence. What we seem to have is the last part of a commentary on the opening passage of Book 2, 328b26-32, in which Aristotle summarises the contents of Book 1.

250. Aristotle is talking about alteration, though this could hardly be gathered from Philoponus' comment.

251. The Greek word translated 'consider', *theôrêsai*, is intransitive and is here followed by a preposition, *peri* ('about'), which would normally take the genitive case. Here, however, Aristotle's text has the accusative.

252. The square brackets are due to Vitelli, whose apparatus reads '*delevi*' (meaning 'I have removed this from the text').

253. One of original uses of the word *stoikheion*, as of its Latin equivalent *elementum* (our 'element'), is to provide a common name for letters of the alphabet.

254. The discomfort with calling earth, air, fire and water 'elements' is that they can be analysed, in Aristotle's view, in terms of pairs of qualities. A similar problem arises in the discussion of Anaxagoras' view earlier, since Anaxagoras thinks the 'elements' are not basic but contain bits of various homoeomeries. In his Clarendon commentary, Williams compares this problem where the literal sense of a term is no longer thought true of its customary referent, to the modern atom: *Aristotle's De Generatione et Corruptione*, p. 152 (SB).

255. This is not an exact quotation of the text of Aristotle, hence the differences in translation of the lemma. Thanks to David Hahm for comments (SB).

256. While *dunamis* is one of the kinds of quality (Philoponus *in Cat.* 134,8ff), some qualities are substantial, *ousiôdeis* (Philoponus *in Cat.* 135,29). Qualities that are *phusikai* can be contrasted with those that are acquired, *epiktêtoi*, and don't complete the substance (Simplicius *in Cat.* 223,31ff.): some powers, like 'rational' for human beings, are considered part of the substance. Philoponus uses the term *phusikê dunamis* elsewhere for the powers that living things have in virtue of their nature (Philoponus *in Cat.* 123,13; 139,4ff.); for the doctrine that the substance of living things depends on its capacities, see *Meteor.* 4.12. See Plotinus' question about the classification of *dunamis* (*Enn.* 6.1.10-11) and the controversy about the status of substantial qualities (e.g. Simplicius *in Cat.* 241,20): Simplicius claims that qualities are *ousiai* in the superlunar realm rather than the sublunar (Simplicius *in Cat.* 241,25). In this passsage Philoponus doesn't seem to contrast innate to acquired powers so much as those of eternal substances like the heavens to those of bodies subject to coming-to-be and perishing. I am grateful to Frans de Haas and John Ellis for references and discussion of this passage (SB).

257. Vitelli suggest Book 7.3 as the passage Philoponus is referring to, but this chapter seems to make no mention of 'natural powers' or 'lives'. It is concerned, however, with things which cannot be said to alter, to 'come to be *something*' as Aristotle calls it in *GC* 1.3, and this is perhaps what Philoponus means in this sentence by 'the passage from not being to being not *simpliciter*'. [On the use of 'lives' as a Neoplatonist term for the powers or different levels of soul, see H.J. Blumenthal, *Aristotle and Neoplatonism in Late Antiquity: Interpretations of the 'De Anima'*, London 1996. Thanks to Tad Brennan for references and discussion (SB).]

258. The term *hupokeimenon* can mean either a substrate or the subject of predication. In the third figure of the syllogism the common term is in subject position (SB).

259. According to Vitelli the MSS have a variety of unhelpful readings, and while he prints between obeli one of these, *ti onta*, he suggests the reading I have translated, *ê panta <hôsper>*.

260. See the companion volume, *Philoponus: On Aristotle On Coming-to-Be and Perishing 1.1-5*, for the criticism of early natural philosophers (SB).]

261. *Timaeus* 50A.

262. *diatattei*. The contrast is between what Plato says in the *Timaeus* passage 49Dff. and what he has ordained elsewhere about the priority of form (SB).

263. The form, e.g., of fire is spelled out by its definition: 'the hot dry'. The contrarieties hot/cold and dry/wet thus occur in the definition of the form of fire, and the element of fire comes to be when matter receives into itself one side of each of these contrarieties, namely, hot and dry.

264. Chs 4-9.

265. Ch. 5.

266. sc. composed of matter and form, not, like the composite bodies just mentioned, composed of other bodies.

267. Literally, 'to one of the things that are in it'.

268. This verb in participle form, *ousiômena*, is coined from the noun *ousia* ('substance'), and presumably means 'made into substances'.

269. Presumably in the sense of 'substance-making'.

270. The Greek word here, *eidopoioi*, would be translated literally into Latin as '*specificientes*', which is a participial form of a verb which survives in English as 'specify'. The Greek could be rendered 'form-making', but to translate it in this way would be to hide from view the development which led to our modern words 'specify', 'specific', etc.

271. *has proslabousa hê hulê ta hapla apotelei sômata*. The verb *apotelei* is related to the term used to describe the view of the medical scientists in 1.8, that forms are 'products' of mixture (SB).

272. *antilêptikai.* The term comes into common use for the ability of sense-organs to apprehend objects in Stoic and Neoplatonist authors: see LSJ, *ad loc.* (SB).

273. Philoponus argues that Aristotle takes perceptibility to be the definitive mark of natural (i.e. physical – the Greek term is *phusikos*) bodies, since perception is an archetypal case of acting and being affected. Aristotle's reason for focusing on touch may be that the qualities definitive of the four sublunary elements – not only the four qualities hot, cold, wet and dry but also heavy and light – are manifestly tangible (SB).

274. *prosekhestatas.* The term implies adherence more than spatial proximity (SB).

275. The participle, *diêkousa,* is relatively rare in Aristotle, but is used in Stoicism. The phrase Philoponus uses here has echoes in the pseudo-Aristotelian *De Mundo* to express the idea of a power pervading everything: 394b11, 396b29, 398b8. See also 165,13 and 182,28 (SB).

276. *diapephoitêkuia,* a usage for which LSJ cites Plotinus *Ennead* 1.1.4 (SB).

277. The best that one can do here is to paraphrase the Greek which is admirably concise: *tôi sunanairein <men> mê sunanairesthai de.* LSJ s.v. *sunanaireô* cite a very un-Berkeleian remark from *Categories* 7b38: *aisthêton anairethen sunanairei tên aisthêsin. hê de aisthêsis to aisthêton ou sunanairei,* as well as (interestingly) Ammonius *in Cat.* 74,6.19. Perhaps there is an echo of the *Categories* passage here. [Aristotle cites as the second kind of priority *to mê antistrephon kata tên tou einai akolouthêsin,* 'not reciprocating in the sequence of their being' (*Cat.* 12, 14a29). Simplicius provides as a gloss on this the phrase *sunanairoun men mê sunanairoumenon de,* the claim that one destroys the other together with it but isn't destroyed with the other. See *Simplicius on Aristotle Categories* 9-15, translated by Richard Gaskin, forthcoming in this series. For a discussion of the view of possibility in the *Categories* passage cited, a view that Aristotle reconsiders, see Richard Sorabji, *Time, Creation and the Continuum,* London and Ithaca N.Y. 1983, pp. 90-1 (SB).]

278. *diaporthmeutikos.* The term here, for ferrying something to and fro, is a Neoplatonist usage: see Philoponus *in DA* 319,15; 329,6ff. (SB).

279. See n. 277 above. The Greek here is *sunanairousai kai mê sunanairoumenai.*

280. The newly discovered Arabic of Alexander shows that Alexander had already ruled out both pairs the same way: Emma Gannagé, 'Alexandre d'Aphrodise *In De generatione et corruptione apud* Gâbir b. Hayyân, K. al-Tasrîf', *Documenti e studi sulla tradizione filosofica medievale* 9 (1998), 35-86, p. 51 (RRKS).

281. Vitelli supplies the word thus translated, which is not in the MSS.

282. See n. 270 above.

283. The Greek is *prôtê,* 'first' or 'primary', but the fact that it is followed by a genitive of comparison shows that Philoponus understands it in the sense of 'prior'.

284. The Greek word thus translated, which normally has a sense more akin to 'subject' has here the sense of 'object'.

285. Again a paraphrase is necessary: see n. 277 above.

286. On the complex interrelationships between different kinds of priority, see Aristotle's *Categories* 12 and *Metaphysics* 5.11 (SB).

287. Philoponus here is trying to contrast the active role implied by the definitions of hot and cold with the passive role implied by the definitions of wet and dry. Aristotle mentioned four qualities, hot and cold, dry and wet: Philoponus stresses that the claim that some are active, others passive shouldn't be understood distributively to mean that hot and dry are the active partner in each pair, but that one pair of contraries are to be viewed as active relative to the other pair (SB).

288. Williams' phrase 'make an absolute distinction' translates *apodiakrinas*. Philoponus wants to recognize that hot and cold are active powers in changes involving composite things, while resisting the implication that in interelemental transformation, one pair of contraries is active and the other passive, as this would suggest that wet and dry were the 'matter' for hot and cold. Philoponus thinks that there is a prime matter that is enformed by the pair of contraries (SB).

289. I use the phrase 'endow with form' to translate *eidopoiountos*. See n. 270 above.

290. Is Philoponus talking about the phenomenon which we should describe as 'having the same melting point'? (Lacking a thermometer it would be difficult for him to connect the possibility of melting with a particular *point*.)

291. See n. 290 above.

292. A medimnus is a corn-measure: we might translate the phrase 'into a bushel of nuts'.

293. The Greeks used the word for 'heavy' to describe sounds that we would describe as 'low' as opposed to 'high-pitched'. Their word for 'high-pitched' is the word that also means 'sharp'.

294. The word translated 'mass' has, of course, none of the theoretical sense the word 'mass' has in modern science.

295. In the sense of 'acute' and 'obtuse'.

296. This is because the word for 'wet' in Greek, *hugros*, also has the sense of 'liquid'. These two senses of *hugros* are what Aristotle is about to make clear.

297. Ch. 9, 10a22.

298. See ch. 2, n. 270.

299. I suggest that *hen ti tôn tessarôn stoikheiôn poiountes* here is an error for *hen ti tôn tessarôn stoikheion poiountes*.

300. The term translated 'operative' here is *dêmiourgika*, which perhaps supplies some precedent in Aristotle for the distinction Philoponus tried to draw earlier between efficient cause and creation. However, Aristotle's usage seems to be closer to Philoponus' *drastêrios* or specifically active cause than to the idea of creation. See nn. 49, 60 and Introduction, above (SB).

301. The Arabic text of Alexander at 330b10 uses this to explain the omission of rare and dense: see Gannagé, 'Alexandre d'Aphrodise *In De generatione et corruptione apud* Gâbir b. Hayyân, K. al-Tasrîf' (RRKS).

302. See Williams' Clarendon commentary, p. 161, on the controversy over the *Divisions*. Philoponus refers to the 'unwritten doctrines' earlier: see n. 105 in the companion volume. For a more recent account of the controversy see, e.g. K. Sayre, *Plato's Late Ontology: A Riddle Resolved*, Princeton 1983 (SB).

303. 'rendered spurious' is Williams' translation of *notheuomena*, a word that is sometimes used to indicate a corrupted text, but is rarely used in this sense. See Galen *de Elem. ex Hipp.*, vol. 1 (Kühn), 454,4. Some of the difficulties with the view that elements are always mixed are discussed in a text attributed to Alexander of Aphrodisias, *Mantissa* 7, forthcoming in this series in a new translation by R.W. Sharples (SB).

304. sc. form and matter.

305. *eilikrines*: see above, Book 1.10, where the term is used of unrestrained functioning. Here the problem is that an Anaxagorean view of the elements would have the same effect as mixing. Rather than talk of restrained capacities, however, the claim is that ubiquitous intermixture renders the elements 'spurious': see n. 303. Presumably the concern is that if an element never occurs in its pure form, then on an Aristotelian account of possibility it is hard to say that it could do so, so that its very nature may seem questionable. Philoponus concludes with the claim that if it is possible and not impeded, a pure form might come about, *ekbaiê an*. Thanks to Bob Sharples for drawing attention to this passage (SB).

306. Timaeus is the character who expounds an elaborate theory of physical reality in Plato's dialogue of that name. Philoponus is referring to what is said in the dialogue.

307. See ch. 2, n. 270.

308. sc. the pair of qualities, hot and dry.

309. This is a puzzling claim. One possible thought behind it may be Aristotle's claim that the 'innate heat', *sumphuton thermon*, of living things is akin to the heat of the fifth element *GA* 2.3, 736b33ff. This heat is found in *pneuma*, the substance of the seed, a substance that the Stoics view as responsible for the cohesion of bodies. But it is difficult to see how Philoponus might attribute this view to Aristotle (SB).

310. sc. their 'natural' places: (1) up, i.e. near the celestial spheres, and (2) down, i.e. near the centre of the earth. [But Alexander takes the reference to be to two qualities, heavy and light, in the Arabic translated by Emma Gannagé: 'Alexandre d'Aphrodise *In De generatione et corruptione apud* Gâbir b. Hayyân, K. al-Tasrîf' (RRKS).]

311. An ugly English expression to translate *rhopê*, the tendency of an object to tilt the balance one way or another.

312. The Greek word I have translated 'each' – strictly speaking it has the sense 'each (or either) of two' – is declinable, so that a distinction can be made between singular and plural. When, as here, it is plural, it has the meaning 'each set (in this case, pair)'.

313. Vitelli follows the Aldine editor in supplying this phrase.

314. See n. 311 above. Later in this sentence I have found it necessary to translate *rhopê* 'tendency to move'.

315. Here Philoponus seems to be finding precedent for the terminology of 'pure' and 'mixed' powers by taking Aristotle's remarks about fire and earth being purer than water and air as referring to the power of motion (SB).

316. Ch. 5, 3b24.

317. One would have thought that 'in the strict sense' each was characterised by a pair of these.

318. The justification for assigning 'wet' to air is the most interesting. Philoponus' argument depends on rejecting the equation of the contraries with phenomenal properties bearing the same name, and focusing instead on the definition of wet as 'easily bounded'. His rejection of the definition of 'wet' as the power to make things wet on the grounds that we haven't yet said what wet is, does not seem to be resisting definition by dispositional properties, but rather to be rejecting a self-referential definition (SB).

319. The words 'water by coldness' are inserted by the Aldine editor, with good reason, it seems.

320. See above, Book 1.8 *passim*, for 'passages'.

321. The Greek printed by Vitelli, *leptotêtês*, must be a misprint for *leptotêtos*.

322. Ch. 6, 304b23ff.

323. sc. Aristotle's remark.

324. This phrase represents the Greek *phaneron*, which Vitelli has supplied from Aristotle's text.

325. I may perhaps be forgiven for repeating here what I wrote in my note on Aristotle's text at this point in my volume in the Clarendon Aristotle Series: 'I have used the word "counterpart" to render the Greek *sumbolon*. *Sumbola* are produced by breaking in two some object, e.g., a bone or a coin, each of a pair of friends or contracting parties, A and B, taking one part, so that at some future date A may prove his association with B by producing his part of the object, which can be seen to be the counterpart of B's. Here Aristotle uses it to refer to the quality that is shared by two "consecutive" (cf. 331b4) simple bodies, e.g., the coldness which is a differentia of both water and earth. The coldness that is in the water has its "counterpart" in the coldness that is in earth. The word *sumbolon* has a distinguished place in the history of ideas, being not only the original of our word "symbol", but the normal word for "creed" in Greek theological writing – hence "Symbolic Theology". Cf. also Plato's *Symposium* 191D in Aristophanes' speech.'

326. i.e. those that follow or precede one another in the cycle: fire, air, water, earth, fire.

327. sc. of qualities, i.e., hot/cold and dry/wet.

328. These words represent the Greek *ekhonta*, which is supplied by Vitelli, following the Aldine editor.

329. sc. the contrary elements.

330. See ch. 2, n. 268.

331. sc. the original pair.

332. The word 'heat', in this sense, does not, of course, have a plural in English. What we have in Greek is the neuter plural of the adjective meaning 'hot'; but translating this, as one might do elsewhere, as 'two hot *things*' will hardly do, because Philoponus (and Aristotle, for that matter) is talking about individual qualities, not the substances that possess them. It is not obvious that this sort of talk about them makes sense. (See my note on Aristotle's text ad loc.)

333. See ch. 2, n. 270.

334. Vitelli indicates a lacuna at this point, but it is not clear why he does so.

335. It seems necessary to suppose that *hen* has fallen out after *men* by haplography at this point in the text.

English-Greek Glossary

a fortiori (argument): *ek tou mallon*
action: *poiêsis*
adhesion: *kollêsis*
affection: *pathos*
aggregation: *sunkrisis*
alter: *alloioun*
alteration: *alloiôsis*
analogous: *analogos*
analogy: *analogia*
analysis, breaking down: *analusis*
assumption: *axiôma*
atom: *atomos*
awareness: *sunaisthêsis*

body: *sôma*
  body in circular motion: *kuklophorêtikon sôma*
  heavenly body: *ouranion sôma*
  mathetical body: *mathêmatikon sôma*
  natural body: *phusikon sôma*
  perceptible body: *aisthêton sôma*

category: *katêgoria*
cause: *aition*
  efficient cause: *kinêtikon, poiêtikon*
  final cause: *telikon*
  motive cause: *poietikon*
change: *metabolê*
circular (argument): *diallêlos*
common usage: *koinê khrêsis*
continuity: *sunekheia*
continuous: *sunekhês*
contrariety: *enantiôsis*
converse (n.): *antistrophon*
counterpart: *sumbolon*

disposition: *hexis*
division: *diairesis*

end: *telos*
explanation: *aitia*
extension: *diastasis, diastêma*

fifth element: *pempton stoikheion*
form: *eidos*
  complete form: *teleion eidos*
  pure form: *eilikrines eidos*
  enmattered form: *enulon eidos*
  proper form: *oikeion eidos*
  substantial form: *ousiôdes eidos*

formula: *logos*
fundamental constituent: *arkhê*
fusion: *sumphusin*

God: *theos*

heavenly bodies: *ourania*
homoeomers: *homoiomerê*
homonymous: *homônumon*
homonymously: *homônumôs*
hyperbaton: *huperbatos*
hypothetical proposition: *sunêmmenon*
hypothetical syllogism: *hupothetikos sullogismos*

imagination: *phantastikon*
impassible: *apathês*
incorporeal: *asômatos*

juxtaposition: *parathesis*

mathematical objects: *mathêmatika*
matter: *hulê*
  corporeal matter: *sômatikê hulê*
  ultimate matter: *eskhatê hulê*
  prime matter: *prosekhê hulê*
  proximate matter: *protê hulê*
  special matter: *idiazousa hulê*
  underlying matter: *hupokeimenê hulê*
metaphor: *metaphora*
mingling: *krasis*
mixing: *mixis*
mixture: *krasis*
movement: *kinêsis*
mutually affect (pass.): *antipaskhein*

nature: *phusis*
natural: *phusikê*

order: *taxis*

part: *morion*
passage: *poros*
per accidens: *kata sumbebêkos*
place: *topos*
plenum: *plêres*
position: *thesis*
potentiality: *dunamis*
power: *dunamis*
primary constituents: *prôtai arkhai*
principle: *arkhê*
privation: *sterêsis*
procurer: *proagôgon*
proof: *deixis*
proximately: *prosekhôs*
putting together, composition: *sunthesis*

quality: *poiotês*
  substantial quality: *ousiôdês poiotês*

range: *platos*
ratio: *logos*
reality: *huparxis*
reciprocally affect (pass.): *antipaskhein*
reciprocally change (pass.): *antikineisthai*
relative: *pros ti*

science: *epistêmê*
segment: *tmêma*
segregation: *diakrisis*
sentence: *lexis*
simpliciter: *haplôs*
solid object: *sustasis*
soul: *psukhê*
speak metaphorically: *katakhrasthai*
specifying/differentiating features: *eidopoioi diaphorai*
speech: *logos*
subject: *hupokeimenon*
substance: *ousia*
substratum: *hupokeimenon*
  primary substratum: *prôton hupokeimenon*
suitability: *epitêdeiotês*
support (of a claim): *pistis*
syllogism: *sullogismos*
symmetry: *summetria*
synecdoche: *sunekdokhê*
synonym: *sunônumon*
syntax: *suntaxis*

thought: *epinoia*
touch: *haphê*
touching: *thixis*
turning: *tropê*

ultimately: *eskhatôs*
unchangeable: *ametablêton*
underlie: *hupokeisthai*
underlying thing: *hupokeimenon*
union: *henôsis*
unlimited: *apeiron*

verbal misapplication: *katakhrêstikôs*
visual ray: *opsis*
void: *kenon*

whole: *holon*

Greek-English Index

aitia, explanation, 154,12
aition, cause
kinêtikon, efficient cause, 127,16
poiêtikon, motive cause, 127,16; efficient cause, 152,28
telikon, final cause, 153,1
akros, extreme, 134,27
alloioun, to produce an alteration, 130,25
alloiôsis, alteration, 190,21; 203,9
ametablêton, unchangeable, 126,28
analogia, analogy, 151,23; 166,26
analogos, analogous, 132,22
analusis, analysis, 161,18
breaking down, 193,33
aniemenos, relaxed, 146,19
antikineisthai, to be reciprocally changed, 137,12
antipaskhein, to be mutually affected, 129,11; to be reciprocally affected, 142,1
antistrephein, to be convertible, 180,22
antistrephonta, (words) which are convertible, 140,1
antistrophon, the converse, 187,7
apathês, impassible, 127,6; 134,10
apeiron, unlimited, 164,17
arkhê, principle, 154,23; fundamental constituent, 155,24
asômatos, incorporeal, 126,3
atomos, atom, 154,3; 166,28
point-like, 170,34
axiôma, assumption, 193,11

deixis, proof, 144,13
dêmiourgia, creation, 169,7
diairesis, division, 155,5
diakrisis, segregation, 187,13
diastasis, extension, 132,25
diastêma, extension, 132,27
dunamis, potentiality, 142,5; 173,10; power, 198,24
drastêrios dunamis, power of affecting things, 149,1
phusikê dunamis, natural power, 206,5

eidopoioi diaphorai, specifying/differentiating features, 225,16
eidos, form, 146,8; 149,21
eilikrines, pure form, 191,27
enhulon, enmattered form, 211,14
oikeion, own, proper form, 191,28
ousiôdes, substantial form, 204,7
teleion, complete form, 191,27
eilikrinês, pure, 96,15
ek tou mallon, a fortiori (argument), 144,18
enantiôsis, contrariety, 143,17; 199,4
epiginesthai, supervene, 169,7
epinoia, thought, 131,17
epistêmê, science, 164,20
epitasis, intensity, 147,7
epiteinomenos, intensified, 146,18
epitêdeiotês, suitability, 169,7; 188,18

haphê, touch, 131,8; 158,28
haplôs, simpliciter, 126,8; 143,26
henôsis, union, 203,9
hexis, disposition, 152,26; 188,20
holon, whole, 126,10
homoiomerê, homoeomers, 169,33; 195,24
homônumôs, homonymously, 130,30
homoula, things that share a common matter, 134,6; 146,7
hulê, matter, 146,8
eskhatê, ultimate matter, 145,28
hupokeimenê, underlying matter, 129,6
idiazousa, special matter, 124,13
prôsekhê, proximate, 145,27
prôtê, prime matter, 124,13; 145,30

sômatikê, corporeal matter, 124,14
huparxis, reality, 202,10
hupokeimenon, subject, 126,28
underlying thing, 128,13
substratum, 139,21; 176,22; 216,2
hupokeisthai, to underlie, 129,14
hupothetikos sullogismos, hypothetical syllogism, 195,9

katakhrêstikôs, incorrectly, 152,18
kata sumbebêkos, per accidens, 142,30
katêgoria, category, 131,3; 151,21
kekolasmenôs, restrained, 188,23
kenon, void, 157,4
kinêsis, movement, 155,2
koinê khrêsis, common usage, 145,16
kollêsis, adhesion, 201,18
krasis, mixture, 149,30; mingling, 187,23

legomena pollakhôs, words used in multiple ways, 130,29; 221,26
lexis, sentence, 130,3
logos, formula, 174,35; 198,1
ratio, 199,17
speech, 205,8

mathêmatika, mathematical objects, 131,15
metabolê, change, 127,14; 186,3
metaphora, metaphor, 152,18
mixis, mixing, 187,20
morion, part, 126,10

opsis, sight, 179,16
visual rays, 153,27
ourania, heavenly bodies, 129,13
ousia, substance, 131,2; 217,26
phusikê, natural substance, 205,30
ousiômena, made into substances, 212,5

parathesis, juxtaposition, 187,24
pathos, affection, 150,9
pempton stoikheion, fifth element, 144,22
phantasia, impression on the imagination, 130,27
phantastikon, imagination, 130,25
phusis, nature, 145,15
pistis, support of a claim, 216,16
platos, range, 146,16
plêres, plenum, 156,20
poiêsis, action, 197,16
poiotês, quality, 129,2
ousiôdes, substantial quality, 212,10
poros, passage, 153,22
proagôgon, procurer, 153,1
pros ti, relative, 139,15
prôtai arkhai, primary and fundamental constituents, 171,27
psukhê, soul, 130,20

skhêma, figure (of a syllogism), 207,2
skhesis, being disposed (towards something), 134,21
sôma, body, 165,11
aisthêton, perceptible body, 211,26
kuklophorêtikon, body whose motion is circular, 129,8
mathêmatikon, mathetical body, 211,26
ouranion, heavenly body, 134,1
phusikon, natural body, 129,28; 131,19
sôreuomenon, that which is piled up, 161,25
sterêsis, privation, 149,21
sunkrisis, aggregation, 160,22; 187,13
sullêptikos tropos, being used once to apply to both, 197,28
sullogismos, syllogism, 135,13
sumbolon, counterpart, 233,5
summetria, symmetry, 169,9
sumphusin, fusion, 217,8
sunaisthêsis, (self-)awareness, 164,24
sunekheia, continuity, 165,11; 189,25
sunekhês, continuous, 133,11
sunekdokhê, synecdoche (figure of speech, taking the part for the whole or the whole for the part), 127,3
sunêmmenon, hypothetical proposition, 154,30; 195,13
sunônumon, synonym (expression used on different occasions with the same sense), 132,9
suntaxis, syntax, 130,3
sunthesis, putting together, 193,12
sustasis, solid object, 158,18

taxis, order, 137,28
telos, end, 152,26

theos, God, 136,34
thesis, position, 131,11
thixis, touching, 137,23
tmêma, segment, branch, 167,30
topos, place, 131,28
tropê, arrangement, 126,27

Subject Index

acting, 186,22; 140,10ff.
  and being affected, 127,14ff.
  and being reciprocally affected, 150,27
  and causing change, 136,16ff.
actuality, pure vs. restrained, 188,25; 203,13
affections, 160,30ff.; 164,25; 168,1ff.; 172,4ff.; 173,29ff.; 212,5ff. 219,7ff.
  tangible, 231,30ff.
agent, 140,20ff.
  primary and proximate, 155,27
agents and patients, 150,4ff.
aggregation and segregation, 127,16; 217,26ff.
Alexander, 135,3; 137,27; 214,23; 222,33; 223,9; 226,18; 232,2; 234,20; 235,32
alter, 136,32
alteration, 128,8; 160,28; 163,14; 190,21; 212,7ff.; 231,29ff.
Anaxagoras, 126,6; 127,19; 192,14
Anaximander, 206,33; 208,9ff.
Anaximenes, 124,21; 206,30
argument from opposites, 221,21ff.
Aristotle
  Cael., 124,28; 134,18; 134,29; 164,5; 164,11; 231,20
  Cat., 132,5; 135,10; 137,30; 150,10; 223,16; 230,4
  GA, 124,26
  Indv. Magn., 164,11
  Meteor., 124,25; 147,15; 217,2
  Phys., 124,8; 126,4; 131,21; 133,9; 134,7; 135,30; 136,10; 138,1; 142,14; 144,20; 149,19; 157,2; 166,34; 178,22; 182,14; 193,14; 198,19; 206,12; 210,25; 211,19
atoms, 158,19ff.; 166,9ff.
  combination of parts of, 171,15ff.
  six arguments against, 164,10ff.
awareness, 164,24

being affected, 140,10ff.
being affected reciprocally, 142,1ff.
body
  atomic, 159,17; 164,10
  divisible ad infinitum, 193,13
  primary, 161,16; arising from matter, 211,4ff.
  principle of, 206,28ff.
  separable, 207,5ff.
  simple, subsisting over and above the four, 207,6ff.
bounded easily/with difficulty, 219,7ff.

cause
  efficient, 136,34; 152,31ff.
  final, 152,20ff.
causing change, 136,23ff.; 150,8ff.
celestial bodies, 137,24; 199,4; 206,4
change, 154,15ff.
  of contraries into one another, 147,32ff.
  of elements into one another, 205,20ff.; 233,3ff.
  in respect of quality, 136,28; place, 136,32
  primary cause of, 150,17
  proximate cause of, 150,22
  reciprocal, 139,23; 151,10
coming-to-be, 138,25ff.; 149,19ff.; 163,10ff.; 164,16ff.; 186,23ff.; 205,13ff.
  and perishing, 189,22ff.
  of composite bodies, 217,3ff.
contact, 129,23; 137,7ff.; 158,27ff.
  in the case of things that are mixed, 200,10ff.
  mutual, 138,8
  senses of, 137,20
continuity and division, 165,8ff.
continuous as divisible ad infinitum, 193,13ff.
contrariety, 143,17ff.; 147,28ff.; 177,17ff.; 189,4ff.; 211,7ff.

elements called, 224,3ff.
specifying taken on by matter, 213,6ff.
tangible through which primary bodies differ, 215,26ff.
counterpart, 233,5ff.

Democritus, 127,6; 127,21; 140,28; 154,2ff.; 156,17ff.; 158,12; 158,28; 160,25; 161,6ff.; 162,13; 163,27; 174,8; 177,27ff.; 186,4; 193,24
destruction, 202,21
differentiating feature, 204,10; specifying, 213,6; 225,16; 228,21; 230,1
Diogenes of Apollonia, 128,12; 206,29
disposition, 152,22f.; suitable, 169,7ff.
divisible as easily bounded, 200,25
division, 155,5ff.; 182,4ff.; 192,20ff.
listing branches of, 167,30ff.
Divisions of Plato, 226,16ff.

elements, 162,6ff.; 204,14ff.
aggregation of, 160,22ff.
analysis as far as surfaces, 210,9ff.
and their places, 229,7ff.
as principle of bodies, 206,28ff.; 227,21
changing into one another, 232,26ff.
coming-to-be of, 160,22ff.
four, 160,20; 161,19
fifth element, 144,22
not the common matter of perceptible bodies, 206,24ff.
purity of, 228,10ff.
real element of the, 225,8
single, 128,11
Empedocles, 124,18; 125,13; 127,6; 127,17; 153,23; 154,9; 159,16; 160,3ff.; 177,33; 207,19; 212,6; 225,20; 227,21

first principles, 211,12ff.; 213,3ff.; 226,3ff.
form, 169,5ff.; 190,18ff.
pure vs. restrained, 192,12-13

God, 136,34; 152,28
growth, 190,13

Heraclitus, 124,22; 206,29
Hippon, 206,30
Hippasus, 206,29
individual, 148,16
indivisibles, 173,28ff.
infinite in number, 164,17ff.
Ion of Chios, 207,19; 227,14

Leucippus, 154,2; 158,12ff.; 160,8; 161,6; 162,12; 178,1

madness, 158,6
mathematical bodies, 131,11; 132,23; 137,23
matter
as principle of bodies, 206,33; 210,19
as subsisting by itself, 208,23ff.
as substratum for the elements, 209,5ff.
common, 129,11; 145,26; 151,6; 204,8; 206,24
form and, 205,11; 209,13ff.
of opposites, 151,33
passivity of, 153,5ff.
prime, 126,3; 145,30
proximate, 146,24ff.
mixing, 127,29; 154,13ff.; 169,6ff.; 187,20ff.; 189,22f.; 192,21ff.; 197,9ff.; 202,12ff.
as union by means of alteration, 203,9ff.
by dissolution of ingredients, 193,1ff.
relative to perception, 192,22ff.
movement, 154,31; 177,2ff.

natural places, 165,31
natural powers, 206,5ff.
nature, 145,15ff.; 156,14; 175,24ff.
acting in vain, 171,35
numerical identity, 148,9ff.

Parmenides, 154,21ff.; 157,9; 157,26; 207,18; 225,10; 226,3; 227,4
particular, 148,2
parts and wholes, 199,16ff.
passages, 153,22f.; 159,18ff.; 174,16; 177,32ff.
patient, 140,20ff.
peculiar property, 221,5
perceptible bodies, 205,30ff.
perceptible differentiating features, 213,23
perception, 153,25ff.; 154,26; 165,1ff.; 168,7ff.; 232,13ff.

evidence of senses, 155,10
perishing, 139,20ff.
piled-up mass, 161,24ff.
Plato
Divisions, 226,16
Phaedo, 128,21
Sophist, 226,19
Theaetetus, 165,8; 168,11
Timaeus, 162,6; 163,27; 185,18; 208,21; 209,5ff.; 225,13; 226,3; 227,4; 232,23
Unwritten Doctrines of, 226,27
plenum, 156,20ff.; 166,11ff.
plurality, 155,11ff.
position, 133,18; 137,22
potentiality, 173,9ff.
and actuality, 182,13ff.; 188,15ff.
third sense, 188,21
power, 135,21; 138,24; 198,13ff.; 217,16
natural powers, 206,9-16
of acting, 140,21, 200,2
of affecting things, 136,1; 140,21; 141,29; 146,29; 148,22; 149,1; 198,20
of being affected, 136,1; 141,30
of secretion, 183,15
of segregating, 192,3-4
primary things, 208,5ff.
putting together, 193,12

quality, 170,9ff.; 191,8ff.
blunted, 198,23
intensified and relaxed, 146,18
produced by mixture vs. supervening, 169,4ff.
substance-making, 212,9
range, 146,19; 148,29; 170,12
rarefaction and condensation, 225,10ff.
relation, 140,1

sciences, 164,20ff.
self-awareness, 165,11f.
shapes, 165,29ff.; 172,22; 190,17
simplicity of form and matter, 228,4ff.
spheres inside the fixed stars, 133,21ff.
substance
one substance contrary to another, 229,24ff.
substantial form, 204,7
substratum, 128,15; 109,18ff.; 202,2ff.
of the senses, 216,2
name of thing that changes from, 209,27ff.
primary, 151,5
receptive of form, 201,20
supervenience, 169,7ff.

Thales, 124,21; 206,30
Timaeus, 228,13
touch, 130,11; 137,11ff. (see also contact)

universal, 147,35

visual rays, 179,7ff.
void, 154,2; 156,21ff.; 166,9ff.; 174,7ff.; 178,5ff.; 180,10ff.

www.ingramcontent.com/pod-product-compliance
Ingram Content Group UK Ltd.
Pitfield, Milton Keynes, MK11 3LW, UK
UKHW020144290926
15471UKWH00004B/686

9 781780 938776

Commercial and Economic Law in the Slovak Republic